DECISION MATHEMATICS

This is one of four Core Texts particularly developed for AEB's Mathematics Core and Options syllabus for A and AS level awards.

This text covers the **Decision Mathematics Core**. The text also covers the material in the AEB Decision Mathematics syllabus (Paper 11, Part A). The optional material required for both syllabuses is combined in a separate publication in this series, called **Option Topics**.

The development of these texts has been coordinated at the *Centre for Innovation in Mathematics Teaching* as part of the

Enterprising Mathematics for AS and A-level Project

in association with the Wessex Project and AEB. It has been partly funded by a grant from the *Leverhulme Trust* and a donation from *British Steel*. The project has been directed by David Burghes and coordinated by Nigel Price

Enquiries regarding this project and further details of the work of the Centre should be addressed to

Margaret Roddick
CIMT
School of Education
University of Exeter
Heavitree Road
EXETER EX1 2LU

CONTENTS

PREFACE

Chapter 1   GRAPHS

1.0   Introduction   1
1.1   The language of graphs   1
1.2   Isomorphism   3
1.3   Walks, trails and paths   7
1.4   Cycles and Eulerian trails   8
1.5   Hamiltonian cycles   9
1.6   Trees   10
1.7   Coloured cubes   13
1.8   Miscellaneous Exercises   15

Chapter 2   TRAVEL PROBLEMS

2.0   Introduction   17
2.1   The shortest path problem   17
2.2   The minimum connector problem   21
2.3   Kruskal's algorithm   22
2.4   Prim's algorithm   25
2.5   The travelling salesman problem   27
2.6   The Chinese postman problem   32
2.7   Local applications   35
2.8   Miscellaneous Exercises   35

Chapter 3   ITERATION

3.0   Introduction   37
3.1   Crossing ladders   38
3.2   Graphical methods   39
3.3   Improving accuracy   42
3.4   Interval bisection   44
3.5   Linear interpolation   48
3.6   Rearrangement methods   51
3.7   Convergence   53
3.8   Newton's method   54
3.9   The ladders again   59
3.10 Miscellaneous Exercises   61

Chapter 4   INEQUALITIES

4.0   Introduction   63
4.1   Fundamentals   64
4.2   Graphs of inequalities   66
4.3   Classical inequalities   69
4.4   Isoperimetric inequalities   73
4.5   Miscellaneous Exercises   77

Chapter 5   LINEAR PROGRAMMING

5.0   Introduction   79
5.1   Formation of linear programming
      problems   80
5.2   Graphical solutions   83
5.3   Simplex method   87
5.4   Simplex tableau   91
5.5   Miscellaneous Exercises   94

Chapter 6   PLANAR GRAPHS

6.0   Introduction   97
6.1   Plane drawings   98
6.2   Bipartite graphs   99
6.3   A planarity algorithm   100
6.4   Kuratowski's theorem   104
6.5   Miscellaneous Exercises   105

Chapter 7   NETWORK FLOWS

7.0   Introduction   107
7.1   Di-graphs   108
7.2   Max flow - min cut   109
7.3   Finding the flow   110
7.4   Labelling flows   111
7.5   Super sources and sinks   113
7.6   Minimum capacities   114
7.7   Miscellaneous Exercises   116

Chapter 8   LOGIC

8.0   Introduction   117
8.1   The nature of logic   118
8.2   Combining propositions   120

Centre for Innovation in Mathematics Teaching
University of Exeter

S

Editor   David Burghes

Assistant   Victor Bryant
Editors   Ron Haydock
                      Nigel Price

Heinemann Educational

Heinemann Educational
a division of Heinemann Educational Books Ltd.
Halley Court, Jordan Hill, Oxford OX2 8EJ

OXFORD   LONDON   EDINBURGH
MADRID   ATHENS   BOLOGNA   PARIS
MELBOURNE   SYDNEY   AUCKLAND   SINGAPORE
TOKYO   IBADAN   NAIROBI   HARARE
GABORONE   PORTSMOUTH NH (USA)

ISBN 0 435 51553 5
First Published 1992
© CIMT, 1992

Typeset by ISCA Press, CIMT, University of Exeter
Printed and bound by The Bath Press, Avon

8.3	Boolean expressions	123
8.4	Compound propositions	125
8.5	What are the implications?	126
8.6	Recognising equivalence	128
8.7	Tautologies and contradictions	129
8.8	The validity of an argument	131
8.9	Miscellaneous Exercises	132

Chapter 9 BOOLEAN ALGEBRA

9.0	Introduction	135
9.1	Combinatorial circuits	135
9.2	When are circuits equivalent?	138
9.3	Switching circuits	139
9.4	Boolean algebra	142
9.5	Boolean functions	144
9.6	Minimisation with NAND gates	147
9.7	Full and half adders	148
9.8	Miscellaneous Exercises	151

Chapter 10 DIFFERENCE EQUATIONS 1

10.0	Introduction	153
10.1	Recursion	154
10.2	Iteration	156
10.3	First order difference equations	158
10.4	Loans	163
10.5	Non-homogenous linear equations	166
10.6	A population problem	167
10.7	Miscellaneous Exercises	170

Chapter 11 DIFFERENCE EQUATIONS 2

11.0	Introduction	173
11.1	General solutions	174
11.2	Equations with equal roots	178
11.3	A model of the economy	181
11.4	Non-homogeneous equations	182
11.5	Generating functions	187
11.6	Extending the method	189
11.7	Miscellaneous Exercises	192

Chapter 12 CRITICAL PATH ANALYSIS

12.0	Introduction	193
12.1	Activity networks	194
12.2	Algorithm for constructing activity networks	196
12.3	Critical path	200
12.4	Miscellaneous Exercises	204

Chapter 13 SCHEDULING

13.0	Introduction	207
13.1	Scheduling	208
13.2	Bin packing	211
13.3	Knapsack problem	215
13.4	Miscellaneous Exercises	219

ANSWERS	221
FURTHER READING	233
INDEX	235

PREFACE

The latter half of the 20th century has seen rapid advances in the development of suitable techniques for solving decision-making problems. A catalyst for these advances has been the revolution in computing. Even the smallest business is able to obtain, relatively cheaply, significant computing power. These two complementary advances have led to a rapid increase of research into **discrete** or **finite** mathematics.

The vast majority of mathematical developments over the past 300 years since Newton have concentrated on advances in our understanding and application of **continuous** mathematics, and it is only recently that parallel advances in **discrete** mathematics have been sought. Much, although not all, of the mathematics in this text is based on developments this century which are still continuing.

The aim of this text is to give comprehensive treatment of the main topics in discrete mathematics, which are often called **Decision Mathematics,** at a level suitable for A Level courses. In particular, it covers all the topics in the new AEB Paper 11 syllabus for Decision Mathematics, excluding Option Topics which are available in a separate publication. It also forms the basis of the syllabus for Decision Mathematics in the new AEB coursework syllabus, based on the Wessex Project initiative and using a 80% Core plus 20% Option framework. This is illustrated in the diagram opposite.

Each wedge constitutes an AS Level. An A Level in Mathematics can be obtained by taking the Foundation Core plus one Option together with any other Core plus one Option.

The available options relevant to Decision Mathematics include

> **Financial Mathematics**
> **Population Modelling**
> **Codes**
> **Simulation**.

Resources for these (and other Options) are published separately. Full details are available from Heinemann Educational.

Full details of AEB syllabuses in Mathematics are available from

AEB, Stag Hill House, GUILDFORD, Surrey. GU2 5XJ

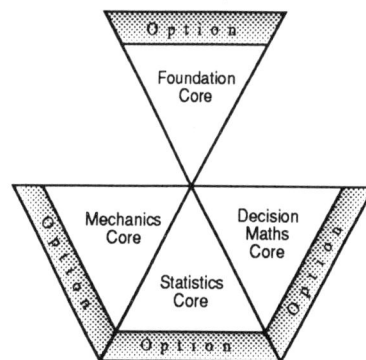

This text has been produced for students and includes examples, activities and exercises. It should be noted that the activities are **not** optional but are an important part of the learning philosophy in which you are expected to take a very active part. The text integrates

- **Exposition**   in which the concept is explained;
- **Examples**   which show how the techniques are used;
- **Activities**   which either introduce new concepts or reinforce techniques;
- **Discussion Points**   which are essentially 'stop and think' points, where discussion with other students and teachers will be helpful;
- **Exercises**   at the end of most sections in order to provide further practice;
- **Miscellaneous Exercises**   at the end of each chapter which provide opportunities for reinforcement of the main points of the chapter.

Discussion points are written in a special typeface as illustrated here.

Note that answers to all the exercises are given at the back of the book. You are expected to have a calculator available throughout your study of this text and occasionally to have access to a computer.

Some of the sections, exercises and questions are marked with an asterix (*). This means that they are either **not** central to the development of the topics in this text and can be omitted without causing problems, or they are regarded as particularly challenging.

The **Teacher's Guide** gives further details and advice on optional material for the AEB syllabus.

ACKNOWLEDGEMENTS

This text has been written by a group of authors working as part of the Mathematics component of the Wessex Project. It has been a delight to work not only with the writers and editors of this text but with the complete team. It says much for their enthusiasm and dedication that we have both a revised AEB syllabus for A Level Mathematics, including the opportunity for two complete A Levels in mathematical subjects and a comprehensive, complete set of texts and resources. In particular, I would like to thank Bob Rainbow (Wessex Project director) for allowing me the opportunity to work with the Wessex project, John Commerford (AEB Mathematics Officer) for his cooperation and patience throughout the development, and Nigel Price for his efficient coordination of the project.

We are also very grateful for the help and assistance of the editorial staff at Heinemann, Philip Ellaway and Ruth Burdett, and to the many organisations which have helped us.

Major funding was provided by the Leverhulme Trust, and we have had generous donations from British Steel, through Colin Green, Tony Nicholson and Brian Taylor. We readily acknowledge their help.

Finally we are all indebted to the support staff, Liz Holland, Margaret Roddick, Ann Tylisczuk and Sally Williams for turning our draft manuscript into an attractive, well-presented text. Their dedication and good humour despite impossible time schedules has been an inspiration to me.

David Burghes
(Project Director)

1 GRAPHS

Objectives

After studying this chapter you should

- be able to use the language of graph theory;
- understand the concept of isomorphism;
- be able to search and count systematically;
- be able to apply graph methods to simple problems.

1.0   Introduction

This chapter introduces the language and basic theory of graphs. These are not graphs drawn on squared paper, such as you met during your GCSE course, but merely sets of points joined by lines. You do not need any previous mathematical knowledge to study this chapter, other than an ability to count and to do very simple arithmetic.

Although graph theory was first explored more than two hundred years ago, it was thought of as little more than a game for mathematicians and was not really taken seriously until the late twentieth century. The growth in computer power, however, led to the realisation that graph theory can be applied to a wide range of industrial and commercial management problems of considerable economic importance.

Some of the applications of graph theory are studied in later chapters of this book. Chapter 2, for example, looks at several different problems involving the planning of 'best' networks or routes, while Chapter 6 considers the question of planarity (very important in designing microchips and other electronic circuits). Chapter 7 deals with problems to do with the flow of vehicles through a road system or oil through a pipe, and Chapters 12 and 13 show how to analyse a complex task and determine the most efficient way in which it can be done. All these applications, however, depend on an understanding of the basic principles of graph theory.

1.1   The language of graphs

A **graph** is defined as consisting of a set of **vertices** (points) and a set of **edges** (straight or curved lines; alternatively called arcs): each edge joins one vertex to another, or starts and ends at the same vertex.

The diagrams show three different graphs, representing respectively the major roads between four towns, the friendships among a group of students, and the molecular structure of acetic acid - the theory of graphs can be applied in many different ways.

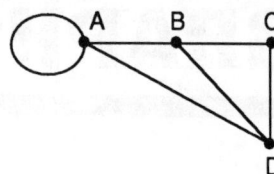

A road system

There are several things to note. One is that although nearly all the edges in these graphs have been drawn as straight lines, this is purely a matter of convenience. Curved lines would have done just as well, because what matters is which vertices are joined, not the shape of the line joining them. Second, each edge joins only two vertices, so that ABC in the first graph is two edges (AB and BC) rather than one long one. Third, the crossing in the middle of the second diagram is not a vertex of the graph; the only points counted as vertices are the ones identified as such at the start.

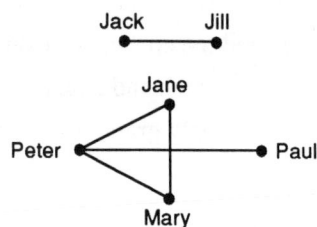

Friendships

The **degree** of a vertex is defined as the number of edges which start or finish at that vertex - an edge which starts and finishes at the same vertex (in other words, a **loop** such as the one at A in the first graph) is counted twice. So, for example, the degree of the vertex A in the first graph is 4, and the degree of the vertex 'Peter' in the second graph is 3. In the third case, the degree of each vertex corresponds to the valency of the atom.

There is actually something a little unusual in the third graph - two edges joining the same two vertices. A multiple edge of this kind can be of great importance in some situations: the difference between saturated and unsaturated fats in a healthy diet, for example, is largely a matter of multiple edges in their molecular structure. In other cases, however, such as the second graph here, a double or triple edge would be meaningless. A graph with no loops and no multiple edges is called a **simple** graph.

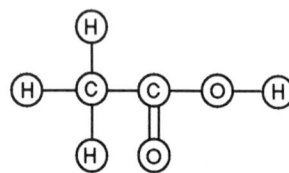

Acetic Acid CH_3COOH

There is an oddity in the second graph too. Jack and Jill are friends with one another but with no one else, so that the graph 'falls apart' into two quite separate pieces. Such a graph is said to be **disconnected**. A **connected** graph is one in which every vertex is linked (by a single edge or a sequence of edges) to every other. If every vertex is linked to every other by a single edge, a simple graph is said to be **complete**.

A **subgraph** of a graph is another graph that can be seen within it; i.e. another graph consisting of some of the original vertices and edges. For example, the graph consisting of vertices 'Jane', 'Mary' and 'Peter' and edges from 'Jane' to 'Mary' and from 'Mary' to 'Peter' is a subgraph of the friendship graph above.

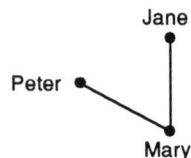

Exercise 1A

Note The answers to these questions will be used in later sections, and should be kept safely until then.

1. For each of the graphs shown below, write down

   (i) its number of vertices,
   (ii) its number of edges,
   (iii) the degree of each vertex.

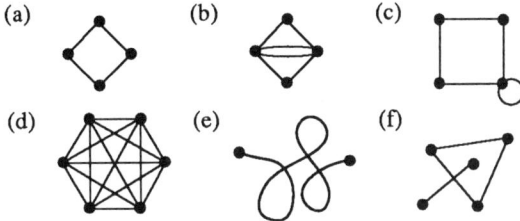

(a)   (b)   (c)

(d)   (e)   (f)

2. Say which (if any) of the graphs in Question 1 are

   (i) simple (ii) connected and/or (iii) complete.

3. Draw graphs to fit the following descriptions:

   (a) The vertices are A, B, C and D; the edges join AB, BC, CD, AD and BD.

   (b) The vertices are P, Q, R, S and T, and there are edges joining PQ, PR, PS and PT.

   (c) The graph has vertices W, X, Y and Z and edges XY, YZ, YZ, ZX and XX.

   (d) The graph has five vertices, each joined by a single edge to every other vertex.

   (e) The graph is a simple connected graph with four vertices and three edges.

1.2 Isomorphism

Look at your answers to Question 3 from Exercise 1A, and compare them with those of other students. You will probably find that some of the drawings look different from others and yet fit the descriptions equally well.

Two graphs which look different, but both of which are correctly drawn from a full description are said to be **isomorphic** - the word comes from Greek words meaning 'the same shape'. Isomorphism is a very important and powerful idea in advanced mathematics - it crops up in many different places - but at heart it is really very simple.

For example, the two graphs shown in the upper diagram each match the full description in Question 3(a) and so are isomorphic to one another. The graphs in the lower diagram each match the description in Question 3(e), but these are not isomorphic. The description did not say which vertices were to be joined by the edges, and the two graphs have joined the vertices differently.

If you are to say that two graphs are isomorphic, there must be a way of labelling or relabelling one or both of them so that the number of edges joining A to B in the first is equal to the number of edges joining A to B in the second, and so on through all possible pairs of vertices. In the upper diagram this is clearly possible: the labelling already on the graphs satisfies this condition, and indeed many people would say that the two graphs are more than isomorphic - they are identical. In the lower diagram, however, no such labelling can ever be found. The second graph has one vertex which is joined to three others, and no labelling of the first graph can ever match this.

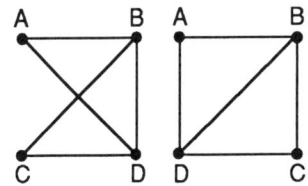

Two possible answers to Question 3 (a)

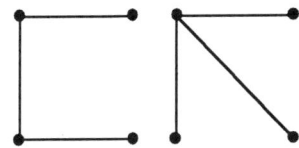

Two possible answers to Question 3 (e)

Testing for isomorphism

If you can match the labels it certainly shows that two graphs are isomorphic, but suppose you cannot. What does that show? It might mean that the graphs are not isomorphic, or it might simply be that you have not yet tried the right labelling combination.

How can you know if the two graphs are isomorphic?

The clue is in the argument that has already been given. If one graph has a vertex of degree three, and the other does not, then no matching can ever be found and the graphs are not isomorphic. This idea can be extended to provide a partial test: a **necessary** condition for two graphs to be isomorphic is that the two graphs have the same number of vertices of degree 0, the same number of vertices of degree 1, and so on. If this condition is not satisfied the graphs are certainly not isomorphic. But it is not a **sufficient** condition; in other words, if the condition is satisfied you still do not know whether or not the graphs are isomorphic and you must go on looking for a match.

Exercise 1B

Look at the graphs below, and say which of them are isomorphic to which others.

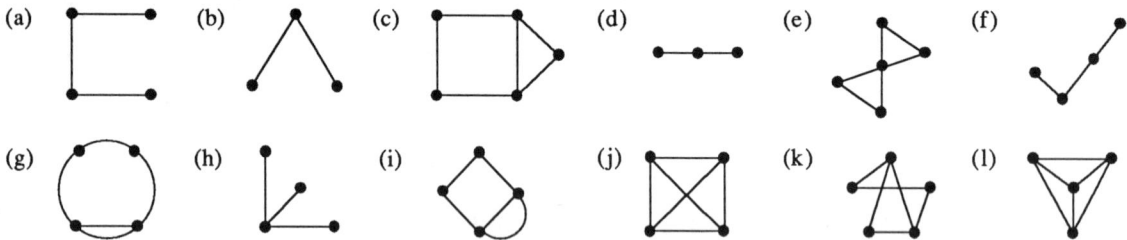

Counting graphs

You may be wondering by now how many different simple graphs can be drawn with just a few vertices. With only one vertex (and no loops allowed) there is clearly only one such graph - the one with no edges.

With two vertices there are two possibilities: there is one graph with no edges and one with one edge, making two possible graphs altogether. You are counting simple graphs, remember, so multiple edges are excluded.

With three vertices there are four possibilities: one each with no edges, one edge, two edges and three edges respectively. Any other simple graph on three vertices must be isomorphic to one of these.

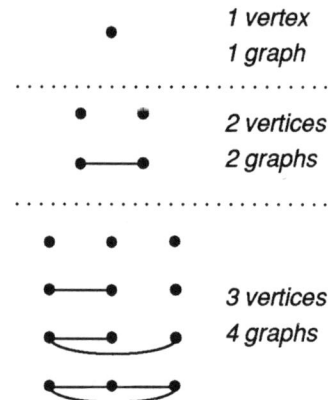

1 vertex
1 graph

2 vertices
2 graphs

3 vertices
4 graphs

*Activity 1   Simple graphs

These results can be put into a table:

Vertices	1	2	3	4	5	6	...
Graphs	1	2	4	...	...	...	...

Try to predict the number of different simple graphs that can be drawn on four vertices, and then check your prediction by drawing them. It is worth doing this in discussion with another student to ensure that you do not leave out any possible graphs nor include two that are actually isomorphic.

When you have got a firm result for four vertices (and corrected your prediction if necessary), try to extend your prediction to five and/or six vertices.

Activity 2   Handshakes

At the beginning of the lesson, greet some of the other members of your group by shaking hands with them. You don't have to shake hands with everyone, and you can shake hands with the same person more than once if you like, but you must keep count of how many handshakes you take part in.

At the end, some members of the group will have been involved in an odd number of handshakes, and others in an even number, so consider this bet: if the number of people involved in an odd number of handshakes is odd, your teacher lets you off homework for a week, but if it is even you get a double dose - does that seem fair?

You may guess that this is not a good bet at all from your point of view - not unless you like doing maths homework, that is! In fact you can never win, because the number of people who shake hands an odd number of times is always even.

Look at your answers to Question 1 in Exercise 1A. Any handshaking situation can be represented by a graph, with people as vertices and handshakes as edges; it may have multiple edges, but not loops. For each graph, find the sum of the degrees of the vertices, and compare it with your other data.

The handshake lemma

You can see at once that the sum of the degrees of the vertices is always twice the number of edges. This is known as the handshake lemma - a lemma is a mini-theorem - and is easy to prove. The degree of each vertex is the number of edge-ends at that vertex, and since each edge has two ends, the number of edge-ends (and hence the total of the vertex degrees) must be twice the number of edges.

This lemma leads quite easily to the unwinnable bet. If the total of individual handshakes is twice the number of handshakes, as the lemma requires, it is certainly an even number. Some members of the class shook hands an even number of times, and the total of any number of even numbers is even. So the total for the rest must be even as well, and since they each shook hands an odd number of times, this can happen only if there are an even number of people. So the number of people involved an an odd number of handshakes must always be even.

The handshake lemma may seem trivial, but it has some quite important consequences and comes up again in Chapter 6.

Activity 3

Try the handshaking exercise again, and this time keep count not of the number of handshakes but of the number of people with whom you shake hands (once or more times makes no difference). What are the chances that at the end there will be two people who have shaken hands with the same number of others?

*The pigeonhole principle

It will perhaps not surprise you to learn that such a coincidence is certain to happen. The proof of this depends on a simple but important principle known as the pigeonhole principle. If n objects have to be put into m pigeonholes, where $n > m$, it says that there must be at least one pigeonhole with more than one object in it. Like the handshake lemma, the pigeonhole principle seems obvious but has a number of uses.

For example, suppose there are 9 people in the room: each must have shaken hands with 0, 1, 2, 3, 4, 5, 6, 7 or 8 others. Of course, if anyone has shaken hands with 8 others - that is, with everyone else - then there cannot be anyone who has shaken hands with 0 others, and vice versa. So among the nine people there are at most eight different scores and the pigeonhole principle says that at least two people must therefore have the same score. You can apply the same argument to any number of people more than one.

* Exercise 1C

These questions are quite challenging, and not the sort of question normally found on an examination paper. If you decide to try them, be prepared to spend quite a lot of time before making progress.

1. By inventing ten appropriate pigeonholes, prove that in any set of eleven whole numbers there are two whose difference divides exactly by 10.

2. Prove that in any set of ten whole numbers between 1 and 20 inclusive, there are two whose highest common factor is greater than 1.

3. By dividing an equilateral triangle of side 2 cm into four appropriate sets, show that if five points are chosen inside the triangle then there will be two of them which are no more than 1 cm apart.

4. A point (x,y) in Cartesian geometry is called a lattice point if x and y are whole numbers. Prove that if any five lattice points are chosen, they include two whose midpoint is also a lattice point.

5. Show that any set of n positive whole numbers includes a non-empty subset (which may be the whole set) whose sum is divisible by n.

1.3 Walks, trails and paths

If you have read any other books about graph theory, you may find this next section rather confusing. Graph theory is a relatively new branch of mathematics, and as yet there is no universal agreement as to the meanings to be given to certain terms. The consequence is that what is called a trail here might be called a walk in another book and a path in a third - the ideas are common but the words are different. The definitions to be used in this book are as follows:

A **walk** is a sequence of edges of a graph such that the second vertex of each edge (except for the last edge) is the first vertex of the next edge. For example, the sequence CD, DA, AB, BD, DA defines a walk (which might be called a walk from C to A) in the graph shown in the diagram. A walk can be the trivial one with no edges at all!

A **trail** is a walk such that no edge is included (in either direction) more than once in the sequence. The walk above is not a trail because the edge DA occurs twice, but CD, DA, AB, BC, CA is a trail from C to A.

A **path** is a trail such that no vertex is visited more than once (except that the first vertex may also be the last); the trail above is not a path because both A and C are visited more than once, but CD, DA, AB is a path from C to B.

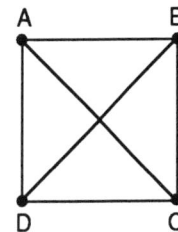

Exercise 1D

1. Referring to the graph in the diagram below, list

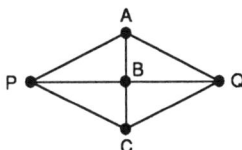

   (a) all the paths from P to Q;
   (b) at least three trails from P to Q which are not paths;
   (c) at least three walks from P to Q which are not trails;
   (d) all the paths which start and finish at P.

2. Which (if any) of the shapes opposite can you draw completely without lifting your pencil from the paper or going over any line twice?

(If you invent appropriate vertices and imagine them as graphs, then you are looking for a trail which includes all the edges)

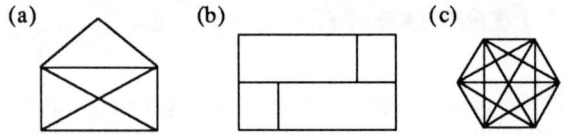

(a)   (b)   (c)

1.4 Cycles and Eulerian trails

Cycles

Puzzles involving trails and paths have been popular for many years, and you may well have seen some or all of the graphs above in books of recreational mathematics. Of particular interest are walks, etc which start and finish at the same place; a walk, trail or path which finishes at its starting point is said to be **closed**, and a closed path with one or more edges is called a **cycle**.

Modern graph theory effectively began with a problem concerning a closed trail. In the 18th century the citizens of the Prussian city of Königsberg (now called Kaliningrad) used to occupy their Sunday afternoons in going for walks. The city stood on the River Pregel and had seven bridges, arranged as shown in the diagram. The citizens' aim was to find a route that would take them just once over each bridge and home again.

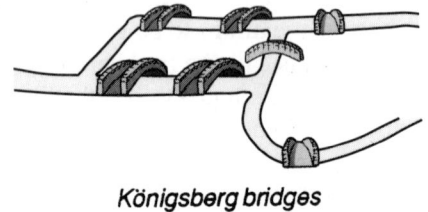

Königsberg bridges

Activity 4   Königsberg bridges

Try to find a route crossing each bridge just once and returning to the starting point.

If you failed, don't worry - so did the people of Königsberg! They began to realise that such a route was impossible, but it was some years before the great Swiss mathematician *Leonhard Euler* proved that this was indeed so. The modern proof, developed from Euler's, is very simple once the bridges are represented by edges of a graph.

How can you be sure that there is no closed trail using all the edges of this graph?

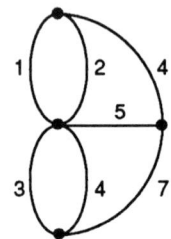

Königsberg bridges in graph form

Eulerian trails

The four vertices of the graph have degrees 3, 3, 3 and 5 respectively - all odd numbers. Any closed trail, on the other hand, goes into a vertex and out of it again, thus adding 2 to its degree on each visit. A closed trail using all the edges cannot exist, therefore, unless every vertex has even degree. (If there are just two vertices with odd degree, they could be the start and finish of a non-closed trail using all the edges.) In fact the opposite is also known: if a connected graph has every vertex of even degree then there does exist a closed trail using all the edges (and if there are just two vertices of odd degree then there is a non-closed trail using all the edges).

As a mark of respect for Euler's work in this area, a trail which includes every edge of a graph is called an **Eulerian trail**. If the trail is closed, the graph itself is said to be **Eulerian**; a semi-Eulerian graph is one that has a non-closed trail including every edge.

Exercise 1E

By considering the degree of each vertex, determine whether each of the graphs shown opposite is Eulerian, semi-Eulerian, or neither. In the case of Eulerian and semi-Eulerian graphs, find an Eulerian trail.

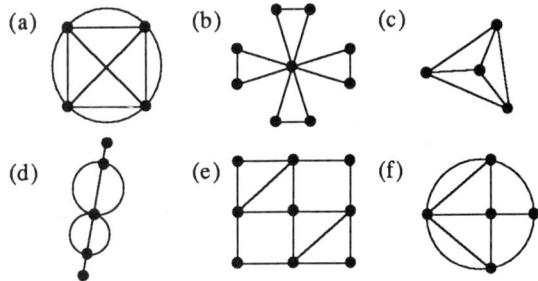

1.5   Hamiltonian cycles

Sir William Rowan Hamilton was a nineteenth-century Irish mathematician who invented in his spare time a game called the **Icosian Game**, based on the vertices of an icosahedron. The idea was essentially simple: given the first five vertices, the player had to find a route that would pass through the remaining fifteen and return to the start without using any vertex twice.

Activity 5   Icosian game

The diagram shows a graph representing a dodecahedron. Try to find such a route - a closed path, to use the modern phrase - beginning with ABCIN in that order.

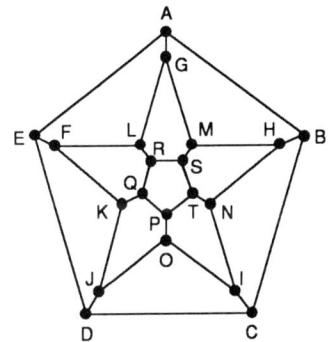

A closed path that passes through every vertex of a graph is called a **Hamiltonian cycle**, and a graph in which a Hamiltonian cycle exists is said to be **Hamiltonian**. The dodecahedron is a Hamiltonian graph, and there are actually two Hamiltonian cycles beginning with the five vertices given:

ABCINHMSTPOJDEFKQRLGA

and   ABCINHMGLFKQRSTPOJDEA.

Distinguishing Hamiltonian from non-Hamiltonian graphs is not easy, and there is no simple test corresponding to the even-degree test for Eulerian graphs.

* Exercise 1F

Decide by trial and error whether or not each of the graphs shown below is Hamiltonian.

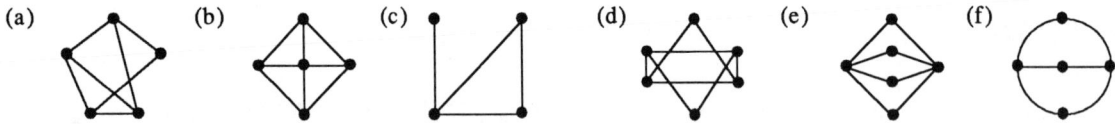

1.6   Trees

A connected graph in which there are no cycles is called a **tree**.

Look at the graphs below and decide which of them are trees.

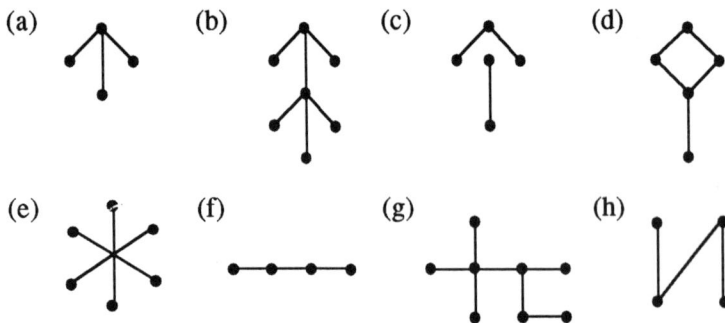

Activity 6

Look again at the graphs you have identified as trees, and count their vertices and their edges. Can you state a general theorem connecting the numbers of vertices and edges for trees? If so, can you prove it?

It is fairly easy to guess from the examples that the number of edges of a tree is always one less than the number of vertices. The proof too is straightforward: if the tree is built up one vertex at a time, starting with one vertex and no edges, each new vertex needs exactly one edge to join it to the body of the tree.

Trees of this kind occur quite often in real life - a biology book may include a 'tree' showing how all living creatures are ultimately descended from the same primitive life forms; a geography text may include a diagram of the entire Amazon river system; and you may find in a history book a diagram showing the line of descent of the Kings and Queens of England, although a certain amount of intermarriage prevents this from being truly a tree as defined above.

Hierarchies

Trees are also commonly used to represent hierarchical organisations. The first diagram below shows part of the management structure of a college, for example, while the second is an extract from a computer's hard disk directory.

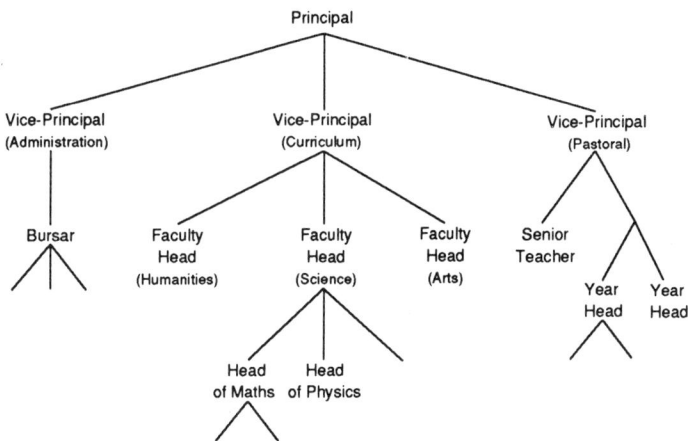

A management Tree

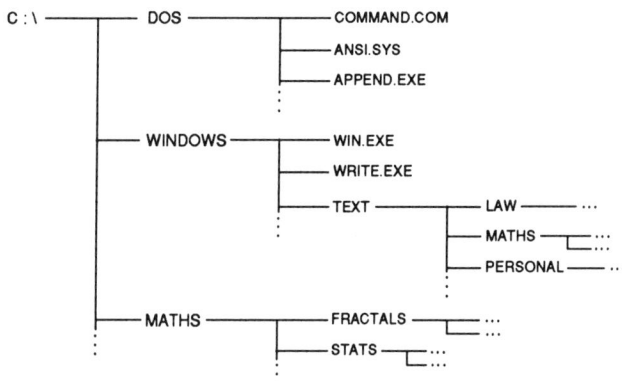

A computer directory tree

Game strategies

Another application of trees is in setting out strategies for playing certain games. For example, the diagram shows the first few stages of a strategy tree for the first player in the game "Noughts and Crosses". You may like to try to complete it.

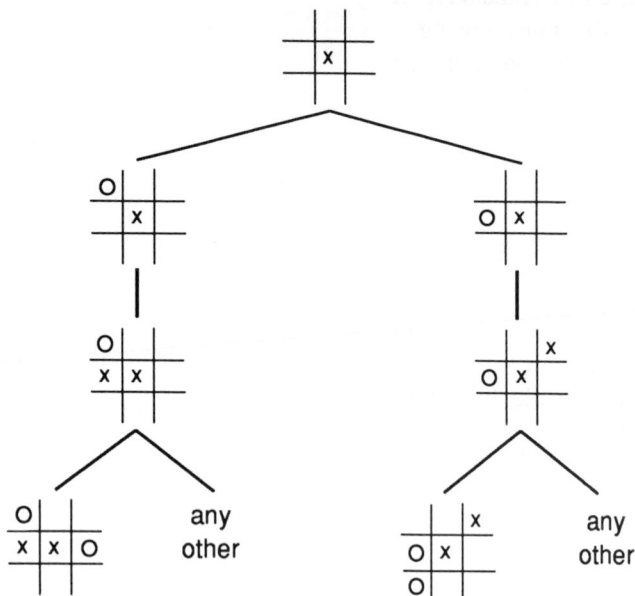

A strategy tree for 'Noughts and Crosses'

Counting trees

There is clearly only one tree with one vertex, one with two, and one with three, as shown in the diagram - any other is isomorphic to one of these. There are two non-isomorphic trees with four vertices, however, and these figures can be set out in a table:

Vertices	1	2	3	4	5	6	7	...
Trees	1	1	1	2	...	...	...	...

	1 vertex
	1 tree

2 vertices
1 tree

3 vertices
1 tree

4 vertices
2 trees

*Activity 7   Counting trees

Draw all the different trees with five vertices, and all those with six, and try to predict from your results the number of seven-vertex trees. Check your prediction by drawing them.

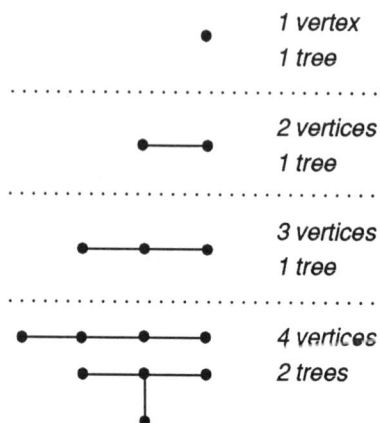

*Organic chemistry

In fact there is no simple formula for unlabelled trees - it turns out to be much easier to count trees if their vertices are labelled - but even the few results in your table can be of use in identifying chemical compounds.

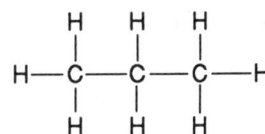
Propane

The group of chemicals known as alkanes have molecules made up of a carbon 'tree' surrounded by hydrogen: each hydrogen atom is bonded to one carbon, and each carbon atom is bonded to four other atoms, of either kind. The diagram shows a molecule of propane (the fuel used in some camping gas stoves), which has three carbon and eight hydrogen atoms and so has molecular formula C_3H_8.

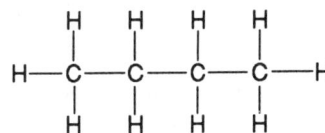
Butane

The molecule can actually be represented completely by its carbon tree, because once all the carbon atoms have been bonded in some formation the hydrogen atoms must go wherever there is a free bond. Now according to the table above there is only one possible carbon tree for CH_4 (methane), only one for C_2H_6 (ethane), and only one for C_3H_8 (propane), so each of these molecular formulae represents only one compound. But there are two distinct trees with four vertices, so the formula C_4H_{10} can represent either butane or isobutane, two different compounds with different properties.

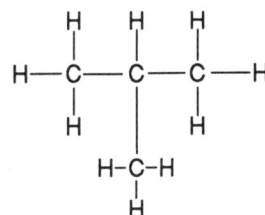
Isobutane
(or 2-methyl-propane)

* Exercise 1G

1. How many different compounds have the molecular formula C_5H_{12}? (If you are studying A Level Chemistry, what are their names?)

2. How many different compounds have molecular formula C_6H_{14}? Think carefully before you answer.

*1.7 Coloured cubes

You may have seen in the shops a puzzle consisting of four cubes with different colours or other designs on their sides. The aim of the puzzle is to stack the cubes in a tower so that each of the long faces shows four different colours or designs. A trial-and-error approach is very difficult, but the application of a little graph theory can lead directly to a solution.

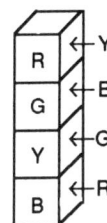

Example

Suppose that the four cubes are coloured as shown .

Cube 1

red	opposite	yellow
green	opposite	yellow
blue	opposite	red

Cube 2

red	opposite	red
green	opposite	blue
blue	opposite	yellow

Cube 3

red	opposite	green
blue	opposite	green
blue	opposite	yellow

Cube 4

red	opposite	yellow
green	opposite	blue
green	opposite	yellow

Transfer all this information to a graph as shown in the diagram, joining the vertices representing opposite colours by an edge numbered to show the cube to which it belongs.

From this graph, extract two disjoint subgraphs - that is, two subgraphs with no edges in common. Each subgraph must consist of four edges of the original graph, chosen in such a way that

(i) the edges include one of each number, in any order,

(ii) each of the vertices R, Y, G, B has degree 2 in each subgraph.

Two subgraphs satisfying these conditions are shown in the lower diagram opposite.

The subgraphs now tell you how to stack the cubes:

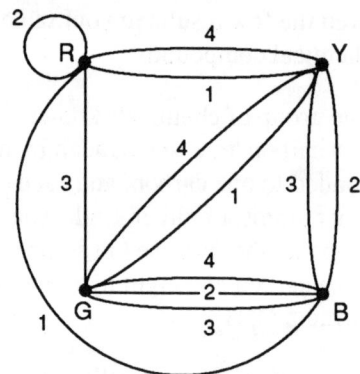

A graphical representation of the four cubes

	From first subgraph		From second subgraph	
	Front	**Back**	**Left**	**Right**
Cube 1	green	yellow	red	blue
Cube 2	red	red	blue	yellow
Cube 3	yellow	blue	green	red
Cube 4	blue	green	yellow	green

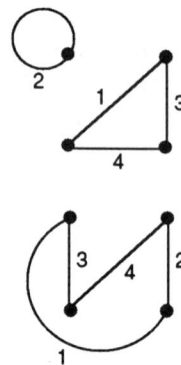

Two disjoint subgraphs

In this particular case there are two other solutions - there is a third subgraph that could have been chosen with either of the two above.

Activity 8

Find the third subgraph and interpret it in the same way as in the example.

Some patterns of cubes have only one solution, however, and others have no solution at all.

*Activity 9

Try to get hold of a commercially-made puzzle of this kind, or make your own, and then amaze your family and friends (and perhaps yourself!) by using mathematics to solve it in just a few minutes.

1.8 Miscellaneous Exercises

1. Draw two simple connected graphs, each with four vertices and four edges, which are not isomorphic.

2. If the vertices of a graph have degree 1, 2, 2, 2 and 3 respectively, how many edges has the graph? Draw two simple connected graphs, each with this vertex set, which are not isomorphic.

3. If P, Q, R, S and T are the vertices of a complete graph, list all the paths from S to T.

4. Determine whether each of the following graphs is Eulerian, semi-Eulerian or neither, and find an Eulerian trail if one exists.

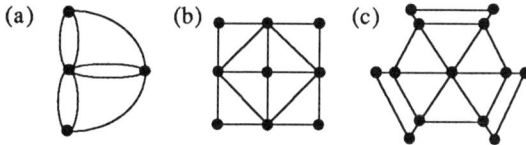

(a)   (b)   (c)

5. You are given nine apparently identical coins, eight of which are genuine, the other being counterfeit and different in weight from the rest - either heavier or lighter, but you do not know which. You are also given a two-sided balance on which to compare the weights of coins or groups of coins. Draw a tree to show a strategy for identifying the counterfeit coin in no more than three weighings.

*6. How many different compounds have molecular formula C_7H_{16}?

*7. Find a Hamiltonian cycle on the graph shown in the diagram.

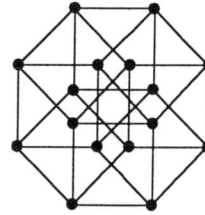

A four-dimensional cube in graph form

*8. Prove that among any group of six people, there are either three who all know one another or three who are mutual strangers.

*9. Given that there are 23 different unlabelled trees with eight vertices, draw as many of them as you can.

*10. A set of four coloured cubes has opposite faces coloured as follows:

Cube 1	R-B, R-Y, B-G;
Cube 2	R-B, Y-Y, Y-G;
Cube 3	R-Y, R-B, B-G;
Cube 4	R-G, G-G, B-Y;

Either find a solution to the four-cube problem or explain why such a solution is impossible.

2 TRAVEL PROBLEMS

Objectives

After studying this chapter you should

- be able to explain the shortest path, minimum connector, travelling salesman and Chinese postman problems and distinguish between them;
- understand the importance of algorithms in solving problems;
- be able to apply a given algorithm.

2.0 Introduction

In this chapter the ideas developed in Chapter 1 are applied to four important classes of problem. The work will make very little sense unless you have studied Chapter 1 already, but no other mathematical knowledge is required beyond some basic arithmetic. A simple calculator may be helpful for some of the work.

2.1 The shortest path problem

The graph opposite represents a group of towns, with the figures being the distances in miles between them. It may seem strange that the direct edge OK is longer than the two edges OH and HK added together, but in real life this is not unusual. You can probably think of examples in your own area where the signposted route from one place to another is longer than an 'indirect' route via a third place, known only to local people.

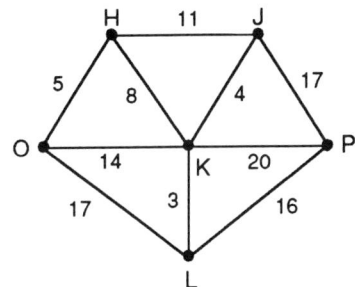

Activity 1   Shortest path

Find by trial and error the shortest route from O to P.

The diagram opposite shows a number of villages in a mountainous area and the time (in minutes) that it takes to walk between them. Once again you may notice that there are cases where two edges of a triangle together are 'shorter' than the third - this is perhaps because the third edge goes directly over the top of a mountain, while the first two go around the side.

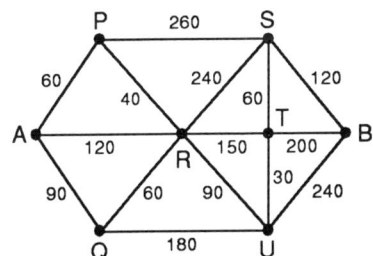

Activity 2   Shortest path

Find the quickest route from A to B.

You may think that there should be some logical and systematic way of finding a shortest route, without relying on a lucky guess. As the number of vertices increases it becomes more and more important to find such a method; and if the problem is to be turned over to a computer, as is usual when problems such as these arise in real life, it is not just important but essential.

What is actually needed is an **algorithm** - that is, a set of step-by-step instructions that can be applied automatically, without any need for personal judgement. A recipe in a cookery book is a good example of an algorithm, as is a well-written set of directions for finding someone's house. The 'Noughts and Crosses' strategy in Section 1.6 is another example, and any computer program depends on an algorithm of some kind.

Activity 3   Finding an algorithm

On your own or in discussion with another student, try to write an algorithm for solving a shortest path problem. That is, write a set of instructions so that someone who knows nothing about mathematics (except how to do simple arithmetic) can solve any such problem just by following your instructions. Test your algorithm on the examples on the previous page.

The shortest path algorithm

The usual algorithm is set out below; you may have come up with something similar yourself, or you may have found a different approach. Study this algorithm anyway, and then decide how you want to handle problems such as these in the future.

Consider two vertex sets: the set S, which contains only the start vertex to begin with, and the set T, which initially contains all the other vertices. As the algorithm proceeds, each vertex in turn will be labelled with a distance and transferred from T to S.

The algorithm runs as follows:

1.   Label the start vertex with distance 0.

2.   Consider all the edges joining a vertex in S to a vertex in T, and calculate for each one the sum of its length and the label on its S-vertex.

3.   Choose the edge with the smallest sum. (If there are two or more with equally small sums, choose any of them at random.)

4.   Label the T-vertex on that edge with the sum, and transfer it from T to S.

5.   Repeat Steps 2 to 4 until the finish vertex has been transferred to S.

6.   Find the shortest path by working backwards from the finish and choosing only those edges whose length is exactly equal to the difference of their vertex labels.

See how the algorithm works on the first example at the beginning of the chapter .

Initially, vertex O is in S, with label 0.

Edges joining {O} to {H, J, K, L, P} are OH $(0+5=5)$, OK $(0+14=14)$ and OL $(0+17=17)$. Choose OH, label H as 5, and transfer H to set S.

Edges joining {O, H} to {J, K, L, P} are OK (14), OL (17), HJ $(5+11=16)$ and HK $(5+8=13)$. Choose HK, label K as 13, and transfer K to set S.

Edges joining {O, H, K} to {J, L, P} are OL (17), HJ (16), KJ $(13+14=17)$, KP $(13+20=33)$ and KL $(113+3=16)$. There are two equal smallest sums, so choose (say) HJ, label J as 16, and transfer J to set S.

Edges joining {O, H, K, J} to {L, P} are OL (17), KL (16), KP (33) and JP $(16+17=33)$. Choose KL, label L as 16, and transfer L to set S.

Edges joining {O, H, K, J, L} to {P} are KP (33), JP (33) and LP $(16+16=32)$. Choose LP, label P as 32, and transfer P to set S.

Working back from P, the edges whose lengths are the difference between their vertex labels are PL, LK, KH and HO (HJ would be a forward, not a backward, move), so the shortest path is O - H - K - L - P, which is 32 miles long.

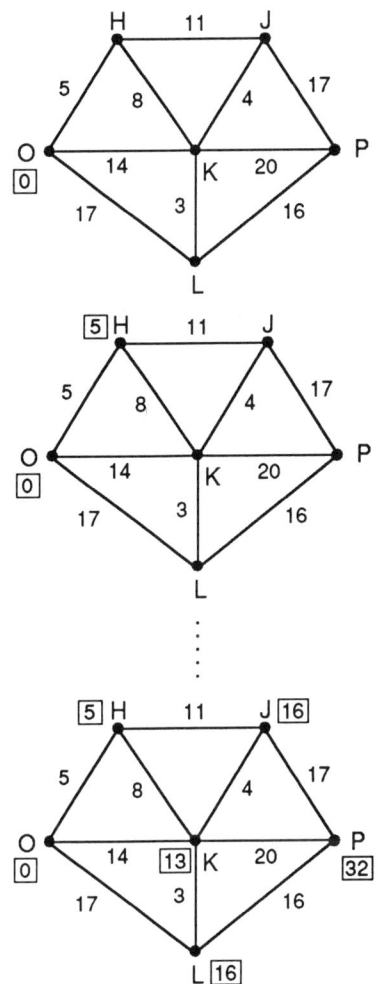

Some stages in the application of the algorithm

The second example can be dealt with similarly. In brief:

> 1. Label A as 0.
>
> 2. Choose AP and label P as 60.
>
> 3. Choose AQ and label Q as 90.
>
> 4. Choose PR and label R as 100.
>
> 5. Choose RU and label U as 190.
>
> 6. Choose UT and label T as 220.
>
> 7. Choose TS and label S as 280.
>
> 8. Choose SB and label B as 400.

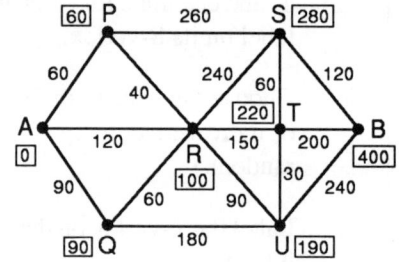

Scan back, choosing BS, ST, TU, UR, RP and PA to obtain the quickest route A - P - R - U - T - S - B, which takes 400 minutes.

Exercise 2A

1. Use the shortest path algorithm to find the shortest path from S to T in each of the diagrams below:

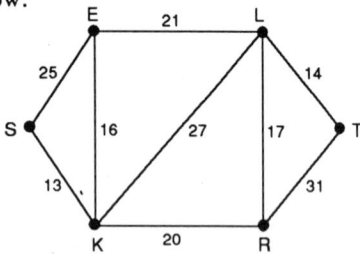

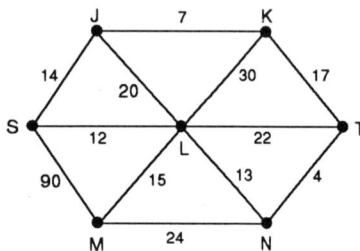

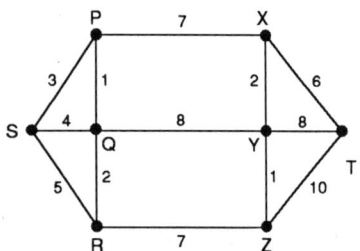

2. The table shows the cost in £ of direct journeys which are possible on public transport between towns A, P, Q, R, S and B. Find the cheapest route by which a traveller can get from A to B.

	A	P	Q	R	S	B
A	-	6	4	-	-	-
P	6	-	-	5	6	-
Q	4	-	-	3	7	-
R	-	5	3	-	-	8
S	-	6	7	-	-	5
B	-	-	-	8	5	-

3. The diagram represents the length in minutes of the rail journeys between various stations. Allowing 10 minutes for each change of trains, find the quickest route from M to N.

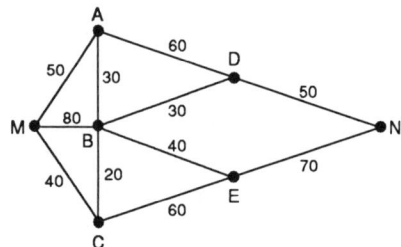

2.2 The minimum connector problem

A cable television company wants to provide a service to each of five towns, and for this purpose the towns must be linked (directly or indirectly) by cable. For reasons of economy, the company are anxious to find the layout that will minimise the length of cable needed.

The diagram shows in the form of a graph - not a map drawn to scale - the distances in miles between the towns. The layout of cable needs to form a connected graph joining up all the original vertices but, for economy, will not have any cycles. Hence we are looking for a subgraph of the one illustrated which is a tree and which uses all the original vertices: such a subgraph is called a **spanning tree**.

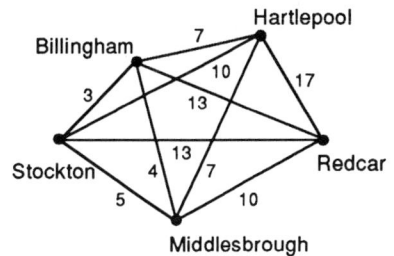

In addition, to find the minimum length of cable needed, from all the possible spanning trees we are looking for one with the total length of its edges as small as possible: this is sometimes known as a **minimum connector**.

The lower diagram shows one of the spanning trees of the graph. It is certainly a tree, and it joins all the vertices using some of the edges of the original graph. Its total length is 26 miles, however, and is not the shortest possible.

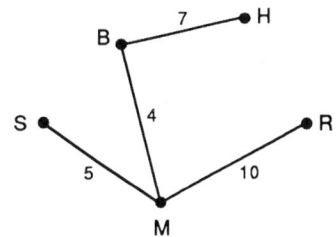

Five towns in Cleveland

Activity 4   Minimum-length spanning tree

Try to find another spanning tree whose total length is minimal.

An oil company has eight oil rigs producing oil from beneath the North Sea, and has to bring the oil through pipes to a terminal on shore. Oil can be piped from one rig to another, and rather than build a separate pipe from each rig to the terminal the company plans to build the pipes in such a way as to minimise their total length.

Activity 5

If the distances in km between the rigs A - H and the terminal T are as shown in the table on the next page, find this minimum total length and say how the connections should be made in order to minimise the total length.

	T	A	B	C	D	E	F	G	H
T	-	120	150	140	120	100	160	70	180
A	120	-	60	60	90	190	210	160	40
B	150	60	-	20	80	180	170	160	50
C	140	60	20	-	40	160	150	140	60
D	120	90	80	40	-	130	70	110	120
E	100	190	180	160	130	-	140	30	220
F	160	210	170	150	70	140	-	150	200
G	70	160	160	140	110	30	150	-	200
H	180	40	50	60	120	220	200	200	-

Like most examples in mathematics books, these problems are oversimplified in various ways. The presentation of the problems implies that junctions can occur only at the vertices already defined, and while this may just possibly be true in the second case (because of the difficulty and expense of maintaining a junction placed on the sea bed away from any rig) it seems unlikely in the first. On the other hand, the terms of the second problem ignore the fact that the costs of laying any pipe depend to some extent on the volume of oil that it is required to carry, and not solely on its length. Even so, these examples serve to illustrate the general principles involved in the solution of real problems.

The first problem above, with just five vertices, can be solved by a combination of lucky guesswork and common sense. The second problem is considerably harder - with nine vertices it is still just about possible to find the minimum connector by careful trial and error, but it is certainly not straightforward. As the number of vertices increases further, so the need for an algorithm grows.

Try to write down your own algorithm for solving problems of this type.

2.3   Kruskal's algorithm

There are two generally-known algorithms for solving the minimum connector problem; your algorithm may be essentially the same as one of these, or it may take a different approach altogether. Whatever the case, work carefully through the next two sections to see how Kruskal's and Prim's algorithms work, and compare the results they give with any that you may have obtained by other methods.

You will recall that the problem is to find a minimum-length spanning tree, and that a spanning tree is a subgraph including all the vertices but (because it is a tree) containing no cycles. **Kruskal's algorithm**, sometimes known as the 'greedy algorithm', makes use of these facts in a fairly obvious way.

The algorithm for *n* vertices is as follows:

> 1. Begin by choosing the shortest edge.
>
> 2. Choose the shortest edge remaining that does not complete a cycle with any of those already chosen. (If there are two or more possibilities, choose any one of them at random.)
>
> 3. Repeat Step 2 until you have chosen *n* – 1 edges altogether; the result is a minimum-length spanning tree.

Example

Look again at the first problem posed in Section 2.2, the diagram for which is repeated here for convenience.

The shortest edge is BS = 3, so choose BS.

The shortest edge remaining is BM = 4, so choose BM.

The shortest edge remaining is SM = 5, but this completes a cycle and so is not allowed. Next shortest is BH or MH, both = 7; so choose randomly (say) BH.

The shortest edge remaining (apart from SM, which is already excluded) is HM, but this would complete a cycle; choose MR as the next shortest.

The four edges now chosen form a spanning tree of total length 24 miles, and this is the solution.

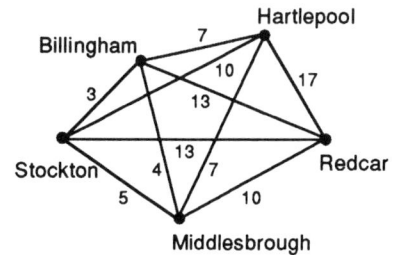

Example

The second problem, for which the distance table is repeated overleaf, can be solved using the same algorithm. You may find it helpful to draw a diagram and mark on the edges as they are chosen. A computer could not do this, of course.

How does a computer identify (and avoid) cycles in applying this algorithm?

	T	A	B	C	D	E	F	G	H
T	-	120	150	140	120	100	160	70	180
A	120	-	60	60	90	190	210	160	40
B	150	60	-	20	80	180	170	160	50
C	140	60	20	-	40	160	150	140	60
D	120	90	80	40	-	130	70	110	120
E	100	190	180	160	130	-	140	30	220
F	160	210	170	150	70	140	-	150	200
G	70	160	160	140	110	30	150	-	200
H	180	40	50	60	120	220	200	200	-

The shortest edge is BC (20), so choose BC.

The shortest remaining is EG (30), so choose EG.

The shortest remaining is AH (40) or CD (40), so choose (say) AH at random.

The shortest remaining is CD (40), so choose CD.

The shortest remaining is BH (50), so choose BH.

The shortest remaining are AB (60), AC (60) and CH (60), but any of these would complete a cycle; the next shortest is TG (70) or FD (70), so choose TG at random.

The shortest remaining (apart from those already excluded) is FD (70), so choose FD.

The shortest remaining is BD (80) but this completes a cycle, as does AD (90); the next shortest is DG (110), so choose DG.

There are nine vertices altogether, so the eight edges now chosen form a minimum-length spanning tree with total length 430 km.

Exercise 2B

1. The owner of a caravan site has caravans positioned as shown in the diagram, with distances in metres between them, and wants to lay on a water supply to each of them. Use Kruskal's algorithm to determine how the caravans should be connected so that the total length of pipe required is a minimum.

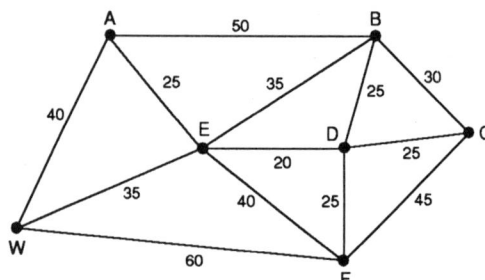

2. The warden of an outdoor studies centre wants to set up a public address system linking all the huts. The distances in metres between the huts are shown in the diagram opposite. How should the huts be linked to minimise the total distance?

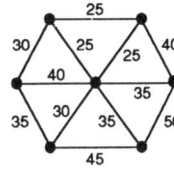

3. A complicated business document, currently written in English, is to be translated into each of the other European Community languages. Because it is harder to find translators for some languages than for others, some translations are more expensive than others; the costs in ECU are as shown in the table opposite.

   Use Kruskal's algorithm to decide which translations should be made so as to obtain a version in each language at minimum total cost.

From / To	Dan	Dut	Eng	Fre	Ger	Gre	Ita	Por	Spa
Danish	-	90	100	120	60	160	120	140	120
Dutch	90	-	70	80	50	130	90	120	80
English	100	70	-	50	60	150	110	150	90
French	120	80	50	-	70	120	70	100	60
German	60	50	60	70	-	120	80	130	80
Greek	160	130	150	120	120	-	100	170	150
Italian	120	90	110	70	80	100	-	110	70
Portuguese	140	120	150	100	130	170	110	-	50
Spanish	120	80	90	60	80	150	70	50	-

2.4 Prim's algorithm

Although Kruskal's algorithm is effective and fairly simple, it does create the need to check for cycles at each stage. This is easy enough when calculations are being done 'by hand' from a graph, but (as you may have discovered) is less easy when working from a table and is quite difficult to build into a computer program.

An alternative algorithm which is marginally harder to set out, but which overcomes this difficulty, is **Prim's algorithm.** Unlike Kruskal's algorithm, which looks for short edges all over the graph, Prim's algorithm starts at one vertex and builds up the spanning tree gradually from there.

The algorithm is as follows:

> 1. Start with any vertex chosen at random, and consider this as a tree.
>
> 2. Look for the shortest edge which joins a vertex on the tree to a vertex not on the tree, and add this to the tree. (If there is more than one such edge, choose any one of them at random.)
>
> 3. Repeat Step 2 until all the vertices of the graph are on the tree; the tree is then a minimum-length spanning tree.

Example

Consider once again the cable television problem from Section 2.2.

For no particular reason, choose Hartlepool as the starting point.

The shortest edge joining {H} to {B, S, M, R} is HB or HM, so choose either of these - HM, say - and add it to the tree.

The shortest edge joining {H, M} to {B, S, R} is MB, so add that to the tree.

The shortest edge joining {H, M, B} to {S, R} is BS, so add that to the tree.

The shortest edge joining {H, M, B, S} to {R} is MR, so add that to the tree.

All the vertices are now on the tree, so it is a spanning tree which (according to the algorithm) is of minimum length.

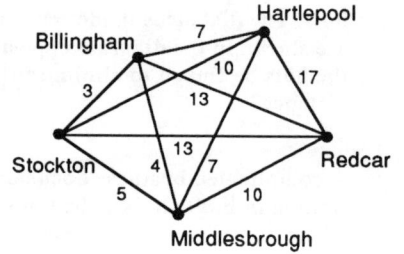

Example

Similarly, consider again the problem about the oil wells at distances as in the table:

	T	A	B	C	D	E	F	G	H
T	-	120	150	140	120	100	160	70	180
A	120	-	60	60	90	190	210	160	40
B	150	60	-	20	80	180	170	160	50
C	140	60	20	-	40	160	150	140	60
D	120	90	80	40	-	130	70	110	120
E	100	190	180	160	130	-	140	30	220
F	160	210	170	150	70	140	-	150	200
G	70	160	160	140	110	30	150	-	200
H	180	40	50	60	120	220	200	200	-

The obvious place to start is the terminal, so place T on the tree.

The shortest edge between {T} and {A, B, C, D, E, F, G, H} is TG, so add this to the tree.

The shortest edge between {T, G} and {A, B, C, D, E, F, H} is GE, so add this.

The shortest edge between {T, G, E} and {A, B, C, D, F, H} is GD.

The shortest edge between {T, G, E, D} and {A, B, C, F, H} is DC.

The shortest edge between {T, G, E, D, C} and {A, B, F, H} is CB.

The shortest edge between {T, G, E, D, C, B} and {A, F, H} is BH.

The shortest edge between {T, G, E, D, C, B, H} and {A, F} is HA.

The shortest edge between {T, G, E, D, C, B, H, A} and {F} is DF.

The tree now includes all the vertices, and thus is the minimum-length spanning tree required.

Exercise 2C

1. Use Prim's algorithm to solve the minimum connector problem for each of the graphs below.

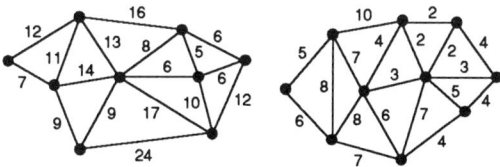

2. A group of friends wants to set up a message system so that any one of them can communicate with any of the others either directly or via others in the group. If their homes are situated as shown below, with distances in metres between them as marked, use Prim's algorithm to decide where they should make the links so that the total length of the system will be as small as possible.

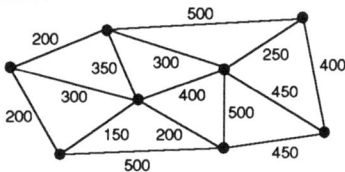

3. The chief greenkeeper of a nine-hole golf course plans to install an automatic sprinkler system using water from the mains at the clubhouse to water the greens. The distances in metres between the greens are as shown in the table below. Use Prim's algorithm to decide where the pipes should be installed to make their total length a minimum.

	CH	1	2	3	4	5	6	7	8
1	250								
2	600	350							
3	400	50	300						
4	100	200	600	300					
5	500	350	200	150	400				
6	800	550	200	500	750	350			
7	600	400	150	300	550	250	150		
8	350	200	350	100	350	50	400	250	
9	50	300	600	400	100	500	850	600	350

[If you want further examples for practice, repeat Exercise 2B using Prim's algorithm instead of Kruskal's algorithm.]

2.5 The travelling salesman problem

A problem not too dissimilar from the minimum connector problem concerns a travelling salesman who has to visit each of a number of towns before returning to his base. For obvious reasons the salesman wants to take the shortest available route between the towns, and the problem is simply to identify such a route.

This can be treated as a problem of finding a minimum-length Hamiltonian cycle, even though in practice the salesman has slightly more flexibility. A Hamiltonian cycle visits every vertex exactly once, you may recall, but there is nothing to stop the salesman passing through the same town twice - even along the same road twice - if that happens to provide the shortest route. This minor difference is rarely of any consequence, however, and for the purpose of this chapter it is ignored.

Activity 6

A salesman based in Harlow in Essex has to visit each of five other towns before returning to his base. If the distances in miles between the towns are as shown on the diagram opposite, in what order should he make his visits so that his total travelling distance is a minimum?

Find a solution by any method, and compare your answer with those of other students.

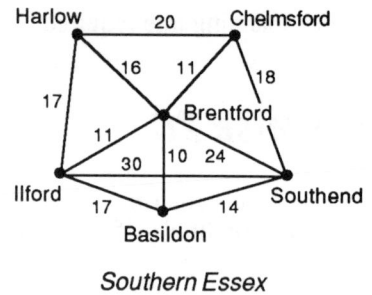

Southern Essex

The shortest route is actually a Hamiltonian cycle in this instance, and has a total length of 90 miles. It can be solved intuitively: there is a fairly obvious 'natural' route around the five outer towns, and all that remains is to decide when to detour via Brentford. For six towns arranged as these are, such a method is generally good enough.

Computer search

An alternative method, impractical for pencil-and-paper calculation but not unreasonable for a computer, is simply to find the total length of every possible route in turn. Given that the route must start and finish at Harlow, there are only 120 different orders in which to visit the other five towns (assuming than no town is visited more than once) and the necessary calculations can be completed in no more than a minute or two. Only half the routes really need to be checked, since the other half are the same routes in reverse, but if the time is available it is much simpler to check them all.

Take a moment or two to think about the procedure that might be used to take the computer through the 120 different routes.

The most obvious way is to keep the first town constant while the computer runs through the 24 possibilities for the other four (which it does by keeping the first of them constant while it goes through the 6 possibilities for the other three, etc.), doing this for each of the five possible first towns. A clever programmer could reduce such a scheme to just a few lines of code using recursive functions - that is, functions defined in terms of themselves!

A less obvious alternative approach is by analogy with bellringing. Traditional English church bellringers do not play tunes, but instead ring 'changes' by ringing the bells in every possible order - essentially the same as the problem here. Many different 'systems' have been devised for ringing changes, with names such as "Plain Bob", "Steadman" and "Grandsire", and any of these can be converted without too much difficulty to a computer program.

Now it is clear that if a computer can be used to check every possible route and find the shortest, that route will be the solution to the salesman's problem. That might appear to be an end to the matter, but in fact it is not. For six towns there are only 120 different routes to try, and that is quite manageable even on a desktop computer, but as the number of towns increases the number of possible routes increases factorially. Thus for ten towns there are 362880 possible routes, for fifteen there are nearly a hundred (US) billion, and for twenty there are more than 10^{17}. Even the fastest computers would take many years, if not many centuries, to check all the possible routes around twenty towns, making the full search method of little use in practice.

Further examples

Example

A business executive based in London has to visit Paris, Brussels and Frankfurt before returning to London. If the journey times in hours are as shown in the diagram opposite, work out the total length of every possible Hamiltonian cycle and thus find the route that takes the shortest time.

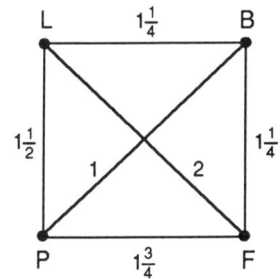

Journey times

There are three cities apart from London, and so 3! = 6 possible orders. The six journeys with their total lengths are as follows:

$$L - P - B - F - L = 5\tfrac{3}{4}\,h \qquad L - F - B - P - L = 5\tfrac{3}{4}\,h$$

$$L - P - F - B - L = 6\tfrac{1}{4}\,h \qquad L - B - F - P - L = 6\tfrac{1}{4}\,h$$

$$L - B - P - F - L = 6\,h \qquad L - F - P - B - L = 6\,h$$

It is clear from this that the shortest route is

London - Paris - Brussels - Frankfurt - London

or the same in reverse.

Example

A visitor to the County Show wants to start from the main gate, visit each of eight exhibitions, and return to the main gate by the shortest possible route. The distances in metres between the exhibitions are given in the table below. What route should the visitor take?

Gate	A	B	C	D	E	F	G	H
Gate	-	200	350	400	500	350	150	200	350
A	200	-	200	300	400	450	300	250	200
B	350	200	-	100	250	450	500	300	100
C	400	300	100	-	150	350	500	300	100
D	500	400	250	150	-	250	450	300	200
E	350	450	450	350	250	-	250	200	350
F	150	300	500	500	450	250	-	200	400
G	200	250	300	300	300	200	200	-	200
H	350	200	100	100	200	350	400	200	-

Without a graph this problem is very difficult, but it is possible to make some sort of attempt at a solution.

Activity 7

Try to find a minimum-length route and compare your answer with those of other students.

The 'best' solution has a total length of 1550 m. This is the length of the route Gate - A - H - B - C - D - E - G - F - Gate, but there are other routes of the same minimum length. If you found any of these routes for yourself, you should feel quite pleased.

You might expect at this point to be given a standard algorithm for the solution of the travelling salesman problem. Unfortunately, no workable general algorithm has yet been discovered - the exhaustive search method is reliable but (for large vertex sets) takes too long to be of any practical use. Intuitive 'common sense' methods can often lead to the best (or nearly best) solution in particular cases, but the general problem has yet to be solved.

* Upper and lower bounds

Although there is no general algorithm for the solution of the travelling salesman problem, it is possible to find upper and lower bounds for the minimum distance required. This can sometimes be very useful, because if you know that the shortest route is between (say) 47 miles and 55 miles long, and you can find a route of length 47 miles, you know that your answer is actually a solution. Alternatively, from a business point of view, if the best route you can find by trial-and-error is 48 miles long, you might well decide that the expense of looking for a shorter route was just not worthwhile.

Finding an upper bound is easy: simply work out the length of any Hamiltonian cycle. Since this cycle gives a possible solution, the best solution must be no longer than this length. If the graph is not too different from an ordinary map drawn to scale, it is usually possible by a sensible choice of route to find an upper bound quite close to the minimum length.

Finding a good lower bound is a little trickier, but not impossible. Suppose that in a graph of 26 vertices, A - Z, you had a minimum length Hamiltonian cycle, AB, BC, CD, ..., YZ, ZA. If you remove any one of its vertices, Z say, and look at the remaining graph on vertices A - Y then

minimum length of a Hamiltonian cycle

=   length of AB, BC, CD, ..., XY, YZ, ZA

=   (length of AB, BC, CD, ..., XY) + (length of YZ)
      + (length of ZA)

$\geq$ $\left(\begin{array}{c}\text{minimum length of a spanning tree}\\\text{of the graph on vertices A - Y}\end{array}\right)+\left(\begin{array}{c}\text{lengths of the two}\\\text{shortest edges from Z}\end{array}\right)$

So a lower bound for the minimum-length Hamiltonian cycle is given by the minimum length of a spanning tree of the original graph without one of its vertices, added to the lengths of the two shortest edges from the remaining vertex.

Look at the County Show problem on the previous page and see how these two methods work. An upper bound is easy to find: the cycle

Gate -A - B -C- D - E - F - G - H - Gate

has total length 1900 m, so the optimum solution must be no more than this. Simply by trial and error, you may be able to find a shorter route giving a better upper bound.

Now let us consider the problem of finding a good lower bound by the method just given. Suppose the vertex E is removed. Using Prim's algorithm as in Section 2.4, the minimum-length spanning tree for the remaining vertices has length 1100 m. The two shortest edges from E are 200 and 250, and adding these to the spanning tree for the rest gives 1550 m. The optimum solution must have at least this length, but may in fact be longer - there is no certainty that a route as short as this exists. The best route around the Show thus has a length between 1550 m and 1900 m.

This technique gives no indication as to how such a best route can be found, of course, although a minimum-length spanning tree may be a useful starter. But if by clever guesswork you can find a route equal in length to the lower bound, you can be certain that it is in fact a minimum-length route. The route Gate - A - H - B - C - D - E - G - F - Gate is an example: it has length 1550 m, equal to the lower bound, and so is certainly a minimum.

Exercise 2D

The following problems may be solved by systematic search by hand or computer, by finding upper and lower bounds, by trial and error, or in any other way.

1. Find a minimum-length Hamiltonian cycle on the graph shown in the diagram below.

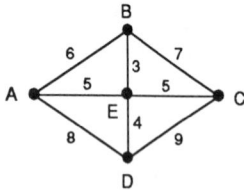

2. The Director of the Scottish Tourist Board, based in Edinburgh, plans a tour of inspection around each of her District Offices, finishing back at her own base. The distances in miles between the offices are shown in the diagram; find a suitable route of minimum length.

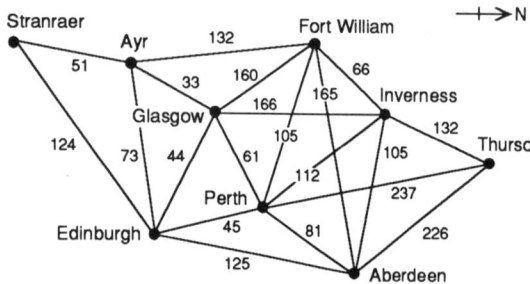

3. A milling machine can produce four different types of component as long as its settings are changed for each type. The times in minutes required to change settings are shown in the table.

From / To	A	B	C	D	Off
A	-	5	7	4	8
B	5	-	6	7	6
C	7	6	-	5	9
D	4	7	5	-	7
Off	8	6	9	7	-

On a particular day, some of each component have to be produced. If the machine must start and finish at "Off", find the order in which the components should be made so that the time wasted in changing settings is as low as possible.

2.6 The Chinese postman problem

The Chinese postman problem takes its name not from the postman's nationality, but from the fact that it was first seriously studied by the Chinese mathematician *Mei-ko Kwan* in the 1960s. It concerns a postman who has to deliver mail to houses along each of the streets in a particular housing estate, and wants to minimise the distance he has to walk.

How does this differ from the problem of the travelling salesman?

The travelling salesman wants to visit each town - each vertex, to use the language of graph theory - so the solution is a Hamiltonian cycle. The postman, on the other hand, wants to travel along each road - each edge of the graph - so his problem can best be solved by an Eulerian trail if such a thing exists. Most graphs are not Eulerian, however, and this is what makes the problem interesting.

The diagram shows a housing estate with the length in metres of each road. The postman's round must begin and end at A, and must take him along each section of road at least once.

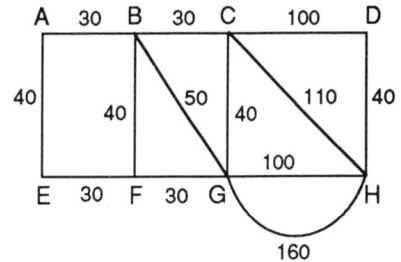

Activity 8

Try to find a route of minimum total length.

Once again, trial and error will often lead to a satisfactory solution where the number of vertices is small, but for larger graphs an algorithm is desirable. A complete algorithm for the solution of the Chinese postman problem does exist, but it is too complicated to set out in full here. What follows is a much simpler partial algorithm that will work reasonably well in most cases.

Systematic solution

The algorithm combines the idea of an Eulerian trail with that of a shortest path. You will recall that an Eulerian graph can be identified by the fact that all its vertices have even degree, and this is at the heart of the systematic solution. The method is as follows:

Find the degree of each vertex of the graph.

1. If all the vertices are of even degree, the graph is Eulerian and any Eulerian trail is an acceptable shortest route.

2. If just two vertices are of odd degree, use the algorithm from Section 2.1 to find the shortest path between them; the postman must walk these edges twice and each of the others once.

3. If more than two vertices are of odd degree - this is where the partial algorithm fails - use common sense to look for the shortest combination of paths between pairs of them. These are the edges that the postman must walk twice.

Example

Consider again the problem above (regarding the layout of roads as a graph with vertices A - H). The only two odd vertices are F and G, and the shortest path between them is obviously the direct edge FG. The postman must therefore walk this section of road twice and all the rest once: a possible route would be

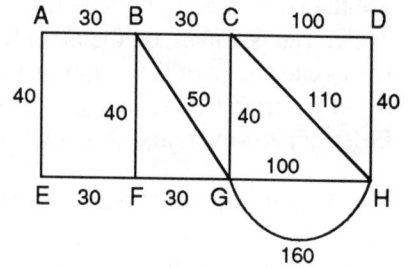

A - B - C - D - H - G - H - C - G - B - F - G - F - E - A.

The total length of any such shortest route is the sum of all the edge lengths plus the repeated edge, which is 830 m.

Example

A roadsweeper has to cover the road system shown in the diagram opposite, going along every road at least once but travelling no further than necessary. What route should it take?

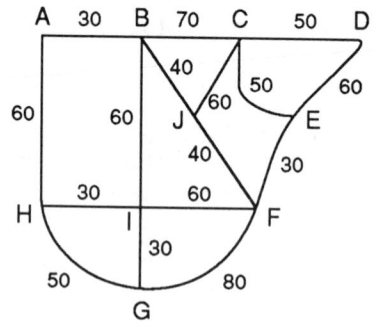

There are four odd vertices, E, G, H and J. By common sense, the shortest combined path comes from joining H to G directly, and J to E via F, so the repeated edges must be HG, FJ and FE with a combined length of 120 m. The edges of the graph have total length 800 m, so the sweeper's best route is 920 m long. One such route is

A - B - C - D - E - C - J - F - E - F - I - B - J - F - G - I - H - G - H - A

but there are many other equally valid routes.

Exercise 2E

1. Find a solution to the Chinese postman problem on each of the graphs opposite.

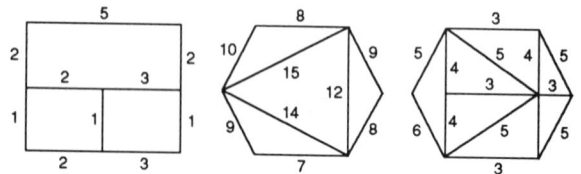

2. A church member has to deliver notices to the houses along each of the roads shown in the diagram below.

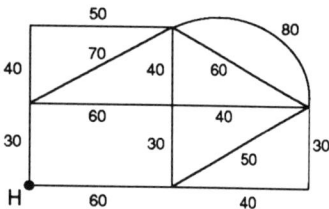

If her own house is at H, what route should she follow in order to make her total walking distance as short as possible?

3. After a night of heavy snow, the County Council sends out its snow plough to clear the main roads shown (with their lengths in km) in the diagram below.

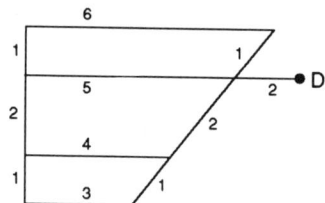

The plough must drive at least once along each of the roads to clear it, but should obviously take the shortest route starting and finishing at the depot D. Which way should it go?

2.7 Local applications

The previous sections of this chapter have covered four classes of problem: the shortest path problem, the minimum connector problem, the travelling salesman problem and the Chinese postman problem. They have some similarities, but each class of problem requires a slightly different method for its solution, and it is important to recognise the different problems when they occur.

One way of acquiring the ability to distinguish the four problems from one another is to consider a selection of real life problems and try to classify them - even try to solve them, if they are not too complicated. Time set aside for this purpose will not be wasted.

The problems you find will depend on your own local environment, but might include

- finding the quickest way of getting from one point in the city to another, on foot, or on a cycle, or by car, or by bus;

- finding the shortest route around your school or college if you have a message to deliver to every classroom;

- finding the shortest route that takes you along every line of your nearest Underground or Metro system, and trying it out in practice;

- planning a hike or expedition visiting each of six pre-chosen churches, or hilltops, or other schools, or pubs;

- whatever catches your imagination.

2.8 Miscellaneous Exercises

1. Use Kruskal's or Prim's algorithm to find a minimum-length spanning tree on the graph below.

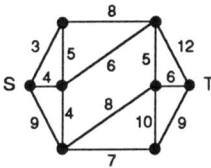

2. Use the algorithm given to find the shortest path from S to T on the graph above.

3. Solve the Chinese postman problem for the graph shown in the diagram above.

4. Find a minimum-length Hamiltonian cycle on the graph above.

5. A railway track inspector wishes to inspect all the tracks shown in the diagram below, starting and finishing at the base B (distances shown in km).

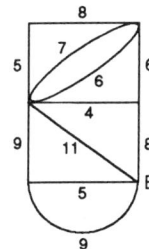

Explain why this cannot be done without going over some sections of track more than once, and find the shortest route the inspector can take.

6. A building society with offices throughout Avon wants to link its branches in a private computer network.

   The distances in miles between the branches are as shown in the table opposite.

   Find a way of connecting them so that the total length of cable required is a minimum.

	BA	BR	CS	CL	KE	KI	LU	PA	PO	RA	WE
Bath		13	12	23	7	9	16	16	22	8	29
Bristol			11	12	6	5	8	7	9	14	21
Chipping Sodbury				23	11	7	19	8	19	20	32
Clevedon					17	17	8	16	5	22	8
Keynsham						4	10	11	15	9	23
Kingswood							13	7	14	13	26
Luisgate								15	8	14	13
Patchway									11	20	24
Portishead										21	13
Radstock											24
Weston Super Mare											

7. A sales rep based in Bristol has to visit shops in each of seven other towns before returning to her base. The distances between the towns are as shown in the diagram below.

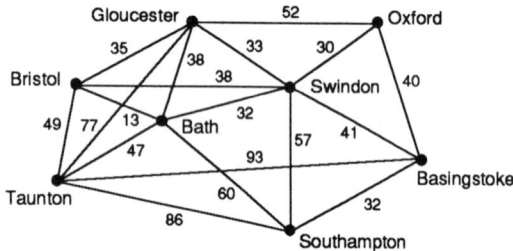

Find a suitable route of minimum length.

8. The diagram below shows the various possible stages of an air journey, each marked with its cost in dollars.

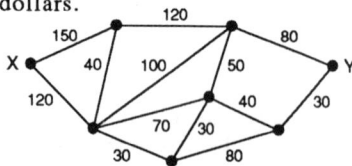

Use a suitable algorithm to find the least expensive route from X to Y, and state its cost.

9. The groundsman of a tennis club has to mark out the court with white lines, with distances in feet as shown in the diagram. Because the painting machine is faulty, it cannot be turned off and so must go only along the lines to be marked. How far must the groundsman walk, given that he need not finish at the same place he started?

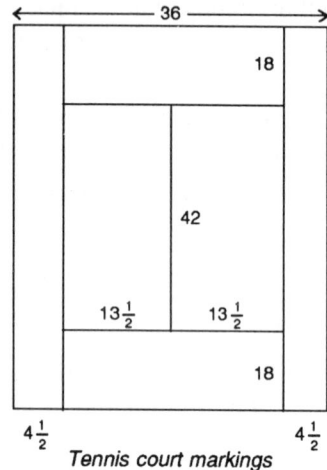

Tennis court markings

	L	O	C	S	G	E	B	N	H
London	-	$1\frac{1}{2}$	$1\frac{1}{2}$	3	5	$4\frac{1}{2}$	2	3	1
Oxford	$1\frac{1}{2}$	-	3	$1\frac{1}{2}$	6	6	2	$4\frac{1}{2}$	$1\frac{1}{2}$
Cambridge	$1\frac{1}{2}$	3	-	4	6	$5\frac{1}{2}$	$3\frac{1}{2}$	2	$2\frac{1}{2}$
Stratford	3	$1\frac{1}{2}$	4	-	5	$6\frac{1}{2}$	$3\frac{1}{2}$	$5\frac{1}{2}$	3
Grasmere	5	6	6	5	-	4	7	8	6
Edinburgh	$4\frac{1}{2}$	6	$5\frac{1}{2}$	$6\frac{1}{2}$	4	-	$6\frac{1}{2}$	7	$5\frac{1}{2}$
Bath	2	2	$3\frac{1}{2}$	$3\frac{1}{2}$	7	$6\frac{1}{2}$	-	5	2
Norwich	3	$4\frac{1}{2}$	2	$5\frac{1}{2}$	8	7	5	-	4
Heathrow	1	$1\frac{1}{2}$	$2\frac{1}{2}$	3	6	$5\frac{1}{2}$	2	4	-

10. An American tourist arrives at Heathrow and wants to visit London, Oxford, Cambridge, Stratford-on-Avon, Grasmere, Edinburgh, Bath, and her ancestors' home in Norwich before returning to Heathrow to catch a flight back to the USA.

    The travel times in hours between places are as shown in the table opposite.

    If she wants to spend 6 hours in each place, can she complete such a journey in the 75 waking hours she has available before her flight leaves?

3 ITERATION

Objectives

After studying this chapter you should

- understand the importance of graphical and numerical methods for the solution of equations;
- understand the principle of iteration;
- appreciate the need for convergence;
- be able to use several iterative methods including Newton's method.

3.0 Introduction

Before you begin studying this chapter you should be familiar with the basic algebraic and graph-plotting techniques covered in higher-level GCSE courses. You should also be able to differentiate at least simple algebraic functions: differentiation is covered in the Foundation Core. A number of examples and exercises involve trigonometry, but these are not essential and can be missed out if you have not covered that topic.

The solution of algebraic equations has always been a significant mathematical problem, and early Egyptian and Babylonian sources show how people of those civilizations were solving **linear, quadratic** and **cubic equations** more than three thousand years ago. The Egyptians often solved linear equations using an 'aha' method (named after the Egyptian word for a heap, not because the answer came as a surprise!) in which they guessed an answer, tried it out, and then adjusted it; the Babylonians solved quadratic and cubic equations by using well-known algorithms together with written-out tables of values.

The spread of Greek mathematics, with its emphasis on elegance and precision, led to the disappearance of those early techniques among academic mathematicians, even if some of them survived among merchants, builders and other practical people. Instead, Mathematicians were more concerned to find general 'analytic' methods based on formulae for the exact solution of any equation of a particular type. Methods of solving quadratic equations were already known, but the first general method for solving a cubic equation was discovered by the Italian mathematician *Scipione del Ferro* in about 1500, and that for **quartics** by his compatriot *Ludovico Ferrari* some fifty years later.

At that point the process of discovery came to a stop, because no one was able to find a method for solving a general equation in x^5 or any higher power. For these equations, as they arose, people had to go back to the earlier trial-and-improvement methods, but the general slowness of those meant that the 'modern' analytic methods were much better. The development of electronic calculators and computers changed all this, however, so that nowadays it is often quicker to use a numerical method such as the Egyptians or Babylonians might have done than to spend time in developing a formula.

You will need a pocket calculator throughout the chapter; a graphic calculator will be particularly useful. Access to a computer with graph-plotting and/or programming facilities may be a further advantage.

3.1   Crossing ladders

In a narrow passage between two walls there are two wooden ladders, a green one 3 m long and a red one 2 m long; the ground is horizontal and the walls are vertical. Each ladder has its foot at the bottom of one wall and its top resting against the other wall. The green ladder slopes up from left to right and the red ladder slopes up from right to left. The ladders cross 1 m above the ground. How wide is the passage?

The first step in solving most problems of this kind is the creation of a mathematical model - not a model made of cardboard and glue, but a set of equations and other relations describing the mathematically important features of the situation.

Stop and think what these important factors are.

The essential facts are summarised in the diagram opposite, which shows the approximate positions and lengths of the ladders and the position of their crossing point. It denotes by x metres the distance to be found, and by u m, v m and w m respectively three other lengths that may be important. The diagram says nothing about the fact that the ladders are made of wood or that they are differently coloured, since these facts are irrelevant to the particular problem to be solved.

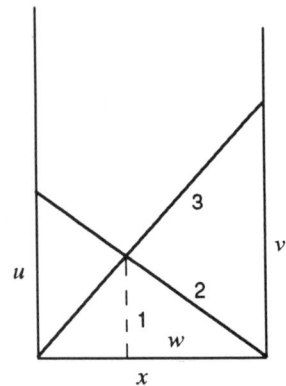

The ladder problem - find the length x

From the diagram, various relationships can be deduced.

Using similar triangles,
$$\frac{w}{1} = \frac{x}{u}$$

Again using similar triangles,
$$\frac{x-w}{1} = \frac{x}{v}$$

Adding these equations,
$$\frac{x}{1} = \frac{x}{u} + \frac{x}{v}$$

$$\Rightarrow \quad \frac{1}{u} + \frac{1}{v} = 1$$

Now by Pythagoras' theorem,
$$u^2 + x^2 = 4$$

$$\Rightarrow \quad u = \sqrt{4 - x^2}$$

and similarly,
$$v^2 + x^2 = 9$$

$$\Rightarrow \quad v = \sqrt{9 - x^2}$$

giving
$$\frac{1}{\sqrt{4 - x^2}} + \frac{1}{\sqrt{9 - x^2}} = 1.$$

All that remains is to solve this equation, and the value obtained for x is the width of the passage in metres.

Consider how you might obtain a solution.

Several methods of solution are possible, but you should have realised almost at once that this is not the sort of equation that can be solved by a simple algorithm (such as, "Take all the x terms to one side and all the numbers to the other"), nor is there a formula such as the one commonly used for quadratic equations. This equation cannot in fact be solved by any such 'analytic' method. Instead we are going to resort to methods which will lead us to very accurate approximations to solutions.

There are two approaches which may work, one involving numerical substitution and the other based on graphs. If you can find a positive numerical value for x which satisfies the equation, then clearly this is a solution. Random guessing is likely to take a long time, however, so any approach must be based on some kind of systematic trial and improvement. Later in this chapter several numerical methods are examined in detail. The alternative graphical approach is the subject of the next section.

3.2 Graphical methods

This section and the next are concerned with the use of graphs of the kind drawn on squared paper rather than those discussed in Chapters 1 and 2. You should already be familiar with the idea of solving an equation by means of a graph: the technique is used in

GCSE work for solving simultaneous linear equations and perhaps quadratic equations too. An example will remind you of the method.

Example

Solve $x^2 + 3x - 5 = 0$.

Solution

The diagram shows the graph of $y = x^2 + 3x - 5$, plotted from a table of values in the usual way. It crosses the x-axis at the points $(-4.2, 0)$ and $(1.2, 0)$ approximately - the graph certainly cannot be read to an accuracy greater than one decimal place - so the solutions of the equation are $x \approx -4.2$ and $x \approx 1.2$.

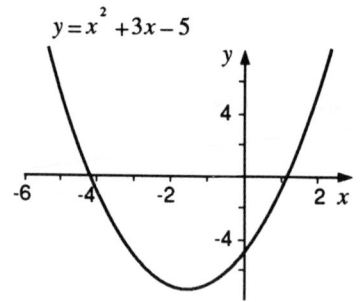

If you have the use of a graphic calculator or a computer with graph-drawing facilities, you can get the same result with much less effort. Draw the graph on the screen, and then use the computer mouse (or the <Trace> function on the calculator) to move the cursor to each of the crossing points in turn.

The same method can be used more generally to solve equations in higher powers of x. There are formulae (like the quadratic formula but much more complicated) for solving cubic and quartic equations, but the French mathematician *Evariste Galois* proved just under 200 years ago that no such formula can ever be found for general equations in powers of x higher than the fourth. As a bonus, the graphical method works for equations including sines, cosines, exponential and logarithmic functions, and so on. Look at some more examples.

Example

Solve the equation $2^x - 3 = 0$.

Solution

The diagram shows the graph of $y = 2^x - 3$, plotted from a table of values or drawn on a calculator or computer screen. The graph crosses the x-axis at $(1.6, 0)$ approximately, and nowhere else, so $x \approx 1.6$ is the only solution of the equation.

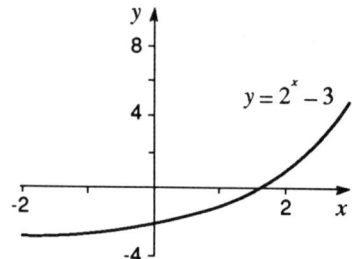

Example

Find correct to one decimal place all the solutions of the equation $5\cos x - x = 0$, where x is expressed in radians.

Solution

The diagram shows the graph of $y = 5\cos x - x$. It crosses the x-axis three times, at $(-3.8, 0)$, $(-2.0, 0)$ and $(1.3, 0)$, and so the equation has the three solutions, $x \approx -3.8$, $x \approx -2.0$, and $x \approx 1.3$.

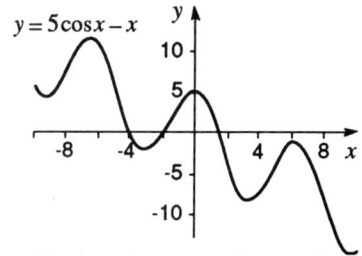

The graph crosses the x-axis three times

Plotting two graphs

The equation in the last example could have been rewritten in the form $5\cos x = x$, so an alternative approach would have been to plot two graphs, the graph of $y = 5\cos x$ and the graph of $y = x$, and to find their points of intersection. The advantage of this method is that both these are well-known functions whose graphs should be familiar, making it quick and easy to draw them. The diagram shows these two graphs, and it is evident that the values of x at the points of intersection correspond to those already found.

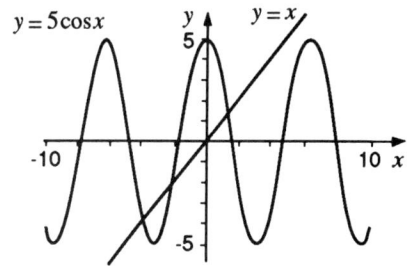

The two graphs intersect at three points

Example

By drawing two graphs, solve $x^3 + 2x - 4 = 0$.

Solution

The equation can be written as $x^3 = 4 - 2x$, and the diagram shows the graphs of $y = x^3$ and $y = 4 - 2x$. They intersect only once, at $(1.2, 1.6)$, so the only solution of this cubic equation is $x \approx 1.2$.

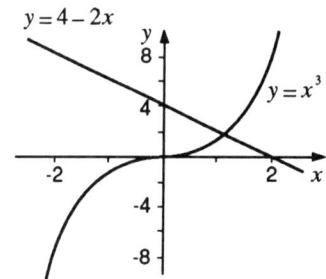

The equation could have been solved in other ways. It could have been rearranged in the form $2x = 4 - x^3$, or even as $x^2 = \dfrac{4}{x} - 2$, and either of these would have given the same result. Often there is no one right way to solve an equation graphically, but a whole collection of ways, some of which may be easier than others.

A possible difficulty

A particular difficulty arises when the graph is nearly flat at the point where it crosses the x-axis, or where two graphs are nearly parallel at their point of intersection. The next example provides an illustration.

Example

Solve the equation $3^x = x^2 + 2x$.

Solution

The first diagram opposite shows the graphs of $y = 3^x$ and $y = x^2 + 2x$, and while it is clear that there is one root of the equation at $x \approx -2.1$, the other (or others?) could be almost anywhere between 1.0 and 2.0.

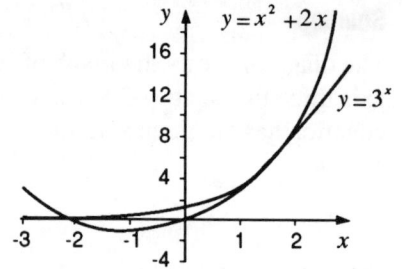

The second diagram showing the graph of $y = 3^x - x^2 - 2x$ is not much more helpful.

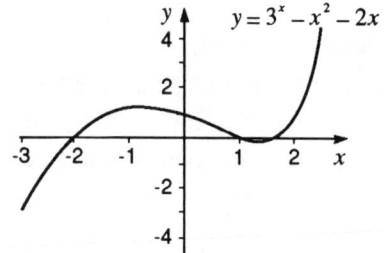

It is only in the third diagram with its exaggerated vertical scale that the other two solutions can be identified as $x \approx 1.0$ ($x = 1$ is actually an exact solution) and $x \approx 1.6$.

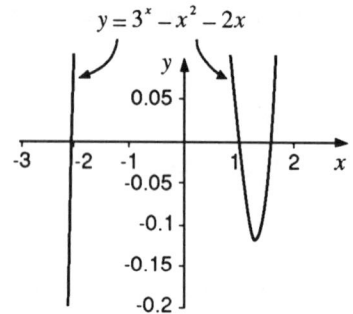

Function on a larger scale

Exercise 3A

Use graphical methods to find approximate solutions of the following equations, giving answers correct to one decimal place.

1. $x^2 - 1 = \dfrac{1}{x}$

2. $x^3 - 6x^2 + 11x - 5 = 0$
3. $x^4 = 2x + 1$
4. $2^x - 5x = 0$
*5. $\sin x = \cos 2x$ (answers between 0 and 2π only)

3.3 Improving accuracy

In the previous section all the solutions were given correct to one decimal place, but this is not always good enough. How might you get a more accurate answer - to two or three decimal places, say?

Stop and think about how you could obtain a more accurate answer.

You may have got a hint from the last example of Section 3.2 - with graphical methods, it is usually possible to get a more accurate answer by redrawing the graph on a larger scale. The next example illustrates this.

Example

Solve $x^3 + 2x^2 - 5 = 0$, giving your answer correct to three decimal places.

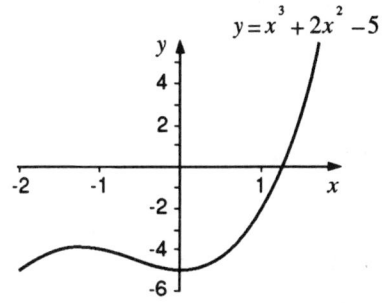

$y = x^3 + 2x^2 - 5$

Solution

The first diagram shows the graph of $y = x^3 + 2x^2 - 5$, plotted for values of x between -2 and 2. The equation clearly has only one root, which lies between 1 and 2, and a graphical estimate might suggest $x \approx 1.2$ to one decimal place.

Graph of function for $-2 \leq x \leq 2$

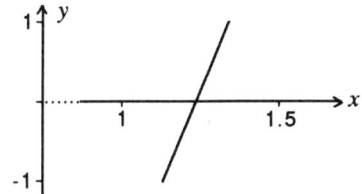

The second diagram shows the same graph, but this time plotted only between $x = 1$ and $x = 1.5$ - notice that over this limited domain the graph is almost a straight line. On this larger scale it is possible to estimate the root more accurately, and to say that $x \approx 1.24$ to two decimal places.

The same graph for $1 \leq x \leq 1.5$

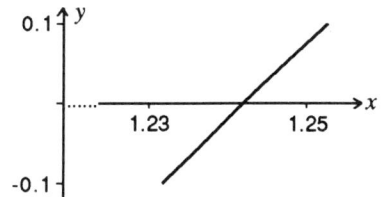

The third diagram increases the scale yet again, and shows the graph plotted over the domain $1.23 \leq x \leq 1.25$. Now the solution can be estimated even more accurately as $x \approx 1.242$ to three decimal places. Clearly there is no limit in theory to the accuracy that can be obtained by this method, but it is time-consuming.

The same again for $1.23 \leq x \leq 1.25$

Computers and calculators

The process can be carried out much more quickly with a computer graph package or a graphic calculator. Most graph-drawing software packages allow the user to change the scales without redrawing the whole graph, and by using this facility to 'zoom in' on the root the solution can be read quite easily to whatever level of accuracy is required.

Activity 1   Graphic calculators

The <Factor> command on the Casio *fx*-7000G is not widely used, but was designed for just this purpose. If you have such a calculator, try the following:

*   <Range> -10, 10, 5, -10, 10, 5
*   <Factor> 5 : <Graph> $Y = Xx^y3 + 2X^2 - 5$ <EXE>
*   <Trace> and use the $\Rightarrow$ and $\Leftarrow$ keys to move the flashing dot close to the point where the graph crosses the *x*-axis, then <EXE> again.
*   Repeat the last instruction as often as necessary - probably

three or four times - until you are satisfied with the accuracy of the x-value given on the screen.

---

*Accuracy

Although in theory there is no limit to the accuracy that can be obtained by a graphical method of this kind, it is difficult in practice to get solutions correct to more than five or six decimal places. Unless you are fond of very cumbersome pencil-and-paper arithmetic you will almost certainly use a calculator to work out the values to plot, and most calculators operate to no more than eight or ten significant figures at best. General-purpose computer packages have a similar limitation - six significant figures is not uncommon - making it impossible to obtain any more accurate result unless you are prepared to adjust the equation as well as the scales.

You should bear in mind too that many equations have **irrational** solutions - that is, the 'true' solutions are numbers that cannot be expressed exactly as fractions or decimals. Thus although it may be possible (in theory) to get as close to the true solution as you might wish, you may never be able to find its exact value. In real life this hardly matters - five or six decimal places is more than enough for any practical purpose - but a mathematician would be careful to distinguish a good decimal approximation from the 'exact' irrational solution.

Exercise 3B

Use a graphical method to solve each of the following equations correct to the stated level of accuracy:

1. $x^3 - 4x + 5 = 0$, to two decimal places

2. $x^5 - x^3 = 1$, to two decimal places

3. $2^x = x + 3$, to two decimal places (both solutions)

4. $2x^2 + 1 = \dfrac{1}{x}$, to three decimal places

*5. $x + \ln x = 0$, to three decimal places

3.4   Interval bisection

The graphical method of solving equations, as you will have realised, has two disadvantages. For one thing it tends to be quite time-consuming, though using a suitable calculator or computer can speed things up considerably. Secondly, however, it needs someone to read the graph, to estimate the position of the crossing point or move the cursor, and (if greater accuracy is required) to decide on the new range of values to be plotted.

These disadvantages are overcome, at least in part, by some of the algebraic methods discussed in the rest of the chapter. The chief benefit of such methods is that they can be expressed algorithmically in terms of yes/no decisions and routine operations, so eliminating the need for human intervention and making them suitable for programming into a computer or calculator.

Activity 2   Guess a number

Take a few minutes to play this game with another student. One of you thinks of a whole number between 1 and 100, and the other has to guess this number by asking no more than ten questions of a yes/no type. Play several rounds, taking it in turns to be the guesser, and try to find the most efficient strategy. How many questions do you really need?

In fact seven questions and a final 'guess' will do, as long as the questions are properly chosen. A skilful guesser might well have asked questions like the following:

Is your number more than 50?	Yes
Is it more than 75?	No
Is it more than 62?	Yes
Is it more than 69?	Yes
Is it more than 72?	No
Is it more than 70?	Yes
Is it more than 71?	Yes
The number is 72.	

At each stage, the guesser is roughly halving the number of possibilities. Initially the number could be anywhere between 1 and 100, but the first answer shows that it is actually between 51 and 100. Then it is between 51 and 75, then between 63 and 75, then between 70 and 75, then between 70 and 72, then between 71 and 72, and the final answer shows that it is 72.

If you played the game enough times you probably discovered this strategy (or something very similar) for yourself. If not, play two or three more rounds using this strategy, to be sure you understand how it works.

Locating a root

The same principle, known as **interval bisection** because at each stage the possible range of values is halved, can be applied to the solution of equations. If you know that a particular equation has a root between 2 and 3 (say), then you can ask whether the root is greater than 2.5, and so halve the interval in which it is to be found. By doing this repeatedly, you can eventually say that the root lies in an interval so small that you can give its value to whatever accuracy you want.

How do you find the first interval (i.e. between 2 and 3) ?

There are at least two practical methods of locating the root, either of which can be used alone but which are much better in combination. The first is to draw a quick rough graph; this does take a little time, but it shows how many roots there are altogether and helps you to avoid any of several possible traps.

The second method, which can be used on its own but which is much more reliable after you have sketched a graph, involves looking for a change of sign. If you find, for example, that $f(2)$ is negative and $f(3)$ positive, or vice versa, it follows that provided f is a continuous function then $f(x)$ is zero somewhere between $x = 2$ and $x = 3$.

The **change-of-sign method** is certainly more precise than the sketch graph in locating a root, but it does contain at least three possible traps. Firstly, it may well locate one root but miss another: unless you are very persevering (or know where to look) you are unlikely to find a root between (say) −11 and −10.

Secondly, the change-of-sign method will not work unless the graph of $y = f(x)$ is continuous over the interval in question. The diagram shows part of the graph of $y = \tan x$ for values of x (in radians) between 0 and 3. It is clear that $f(1) > 0$ and $f(2) < 0$, but equally clear that there is no root of the equation $f(x) = 0$ between 1 and 2.

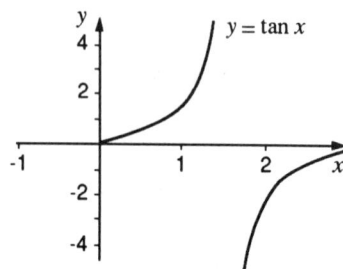

$f(1) < 0$ and $f(2) < 2$
but there is no root between
1 and 2

Thirdly, the change-of-sign method will not show up repeated roots (where the graph just touches the x-axis without crossing it), nor two roots close together. For example, the equation $6x^2 - 29x + 35 = 0$ has solutions $x = 2\frac{1}{3}$ and $x = 2\frac{1}{2}$, as you can check by factorising, but $f(2)$ and $f(3)$ are both positive and so give no indication that these roots exist.

In spite of these three problems, the change-of-sign method of locating roots is very important, and useful too when properly applied in conjunction with a sketch. It forms the basis of the interval bisection and linear interpolation methods discussed in this section and the next, and is commonly used at least at the start of the more sophisticated methods of solution considered later in the chapter.

The method in practice

Example
Solve $x^4 - 2x^3 - 1 = 0$, correct to two decimal places.

Solution
The diagram shows the graph of $y = x^4 - 2x^3 - 1$, and there are evidently two roots, one negative and close to -1, and the other positive and close to 2.

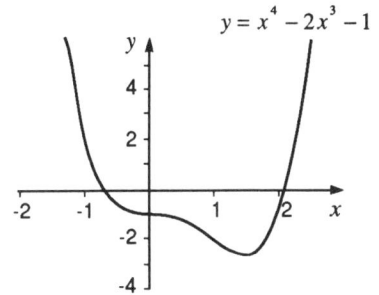

$y = x^4 - 2x^3 - 1$

$f(-1) = 2$, $f(0) = -1$ and the change of sign shows that there is a root between -1 and 0.

$f(2) = -1$, $f(3) = 26$ and the change of sign shows that the second root lies between 2 and 3.

Consider the positive root first.

$f(2.5) \approx 6.8$ so the change of sign is between 2 and 2.5.

$f(2.2) \approx 1.1$ so the change is between 2 and 2.2.

(This is not exactly the midpoint of the interval, but is near enough and keeps the calculation fairly simple.)

$f(2.1) \approx -0.1$ so the change is between 2.1 and 2.2.

$f(2.15) \approx 0.5$ so the change is between 2.1 and 2.15.

$f(2.12) \approx 0.1$ so the change is between 2.1 and 2.12.

$f(2.11) \approx 0.03$ so the change is between 2.1 and 2.11.

$f(2.105) \approx -0.02$ so the change is between 2.105 and 2.11.

This root is therefore 2.11 to two decimal places.

Similarly with the negative root,

$f(-0.5) \approx -0.7$ so the change is between -1 and -0.5.

$f(-0.7) \approx -0.1$ so the change is between -1 and -0.7.

$f(-0.85) \approx 0.8$ so the change is between -0.85 and -0.7.

$f(-0.78) \approx 0.3$ so the change is between -0.78 and -0.7.

$f(-0.74) \approx 0.1$ so the change is between -0.74 and -0.7.

$f(-0.72) \approx 0.02$ so the change is between -0.72 and -0.7.

$f(-0.71) \approx -0.03$ so the change is between -0.72 and -0.71.

$f(-0.715) \approx -0.01$ so the change is between -0.72 and -0.715.

So this root is -0.72 to two decimal places.

Exercise 3C

Use a sketch graph followed by a change-of-sign search to locate the roots of the equation $2^x - 2x - 3 = 0$. Then use an interval bisection method to find these solutions correct to two decimal places.

3.5   Linear interpolation

You may feel that the interval bisection method is unnecessarily slow. If $f(2) = -1$ and $f(3) = 26$, as in the last example with $f(x) = x^4 - 2x^3 - 1$, it is surely obvious that the root is likely to be closer to 2 than to 3. It is as a response to this very reasonable argument that the **linear interpolation** method has been developed.

You saw earlier how any graph drawn on a sufficiently large scale often looks very similar to a straight line, and the linear interpolation method makes use of this fact. The diagram shows a straight line drawn between the points $(2, -1)$ and $(3, 26)$. Since the line cuts the x-axis between 2 and 3 in the ratio $1 : 26$, it must therefore cross the x-axis when $x = 2\frac{1}{27}$, which is approximately the point $(2.04, 0)$. The second approximation produced by linear interpolation is therefore $x \approx 2.04$.

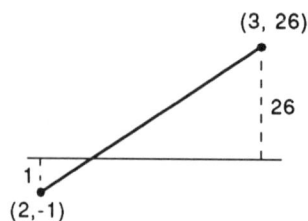

The principle of linear interpolation (not to scale)

Now $f(2.04) \approx -0.66$, and a new straight line can be drawn between $(2.04, -0.66)$ and $(3, 26)$. This divides the x-axis between 2.04 and 3 in the ratio $0.66 : 26$, and therefore cuts it at $(2.06, 0)$.

$f(2.06) \approx -0.48$, and the line between $(2.06, -0.48)$ and $(3, 26)$ cuts the x-axis at $(2.08, 0)$.

$f(2.08) \approx -0.28$, and the line between $(2.08, -0.28)$ and $(3, 26)$ cuts the x-axis at $(2.09, 0)$.

$f(2.09) \approx -0.18$, and the line between $(2.09, -0.18)$ and $(3, 26)$ cuts the x-axis at $(2.10, 0)$.

$f(2.10) \approx -0.07$, and the line between $(2.10, -0.07)$ and $(3, 26)$ cuts the x-axis at $(2.102, 0)$.

After six interpolations, it seems fairly clear that the value is approaching a limit close to 2.10 or 2.11, and it is easy enough to check that $f(2.105) < 0$ and $f(2.115) > 0$, confirming the solution $x \approx 2.11$ to two decimal places. The negative root could be found in a similar way.

Consider a second example.

Example

Find the positive root of $10^x = x + 5$.

Solution

Rearrange the equation in the form $10^x - x - 5 = 0$, and write $f(x) = 10^x - x - 5$. The graph suggests a root between 0 and 1, and this is confirmed by $f(0) = -4$, $f(1) = 4$.

The line between $(0, -0.4)$ and $(1, 4)$ cuts the x-axis at $(0.5, 0)$.

$f(0.5) \approx -2.34$, and the line between $(0.5, -2.34)$ and $(1, 4)$ cuts the x-axis at $(0.68, 0)$.

$f(0.68) \approx -0.89$, and the line between $(0.68, -0.89)$ and $(1, 4)$ cuts the x-axis at $(0.738, 0)$.

$f(0.738) \approx -0.27$, and the line between $(0.738, -0.27)$ and $(1, 4)$ cuts the x-axis at $(0.755, 0)$.

$f(0.755) \approx -0.07$, and the line between $(0.755, -0.07)$ and $(1, 4)$ cuts the x-axis at $(0.759, 0)$.

At this point you may guess that $x \approx 0.76$ is the solution correct to two decimal places, and since you know already that $f(0.755) < 0$, it remains only to check that $f(0.765) > 0$. This is indeed the case, so $x \approx 0.76$ to this level of accuracy.

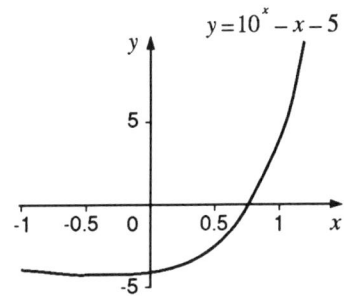

Activity 3

Use a sketch graph followed by a change-of-sign search to locate the roots of the equation $x^3 - 4x^2 + 7 = 0$. Use linear interpolation to find each of these roots correct to two decimal places.

*Interpolation using tables

Although **linear interpolation** is not often used in practice for
solving equations, it is very important in obtaining intermediate
values from printed tables. The table below, for example, is part of
a table showing the monthly repayment required on a 25-year
mortgage of £50 000 at various interest rates.

Rate	8%	9%	10%	11%	12%	13%
Payment	390.33	424.20	459.03	494.75	531.25	568.44

If you plot the rates and repayments accurately on a graph you will
see that they do not lie exactly on a straight line, but they are so
nearly so that linear interpolation can be used to find intermediate
values correct to within a few pence.

Example

Find the monthly repayment if the interest rate is 10.25%.

Solution

10.25 lies between 10 and 11, so the answer lies between 459.03
and 494.75. More specifically, 10.25 divides the interval (10, 11)
in the ratio 0.25 : 0.75, and the point dividing the interval
(459.03, 494.75) in the same ratio is given by

$$\frac{0.25}{(0.25+0.75)} \times (494.75 - 459.03) + 459.03 \approx 467.96$$

So the repayment required will be approximately £468.

Example

Find the interest rate for which the monthly repayment would be
£400.

Solution

400 lies between 390.33 and 424.20, and divides that interval in
the ratio 9.67 : 24.2. The point dividing the interval (8, 9) in the
same ratio is given by

$$\frac{9.67}{(9.67+24.2)} \times (9-8) + 8 \approx 8.29$$

So the interest rate will be about 8.3%.

Many other tables encountered in finance, statistics, science or
engineering call for similar techniques.

* Exercise 3D

Use the table on the previous page to estimate

1. the monthly repayment (to the nearest 10p) if the interest rate is 12.6%;

2. the interest rate (correct to 0.1%) corresponding to a repayment of £500.

3.6   Rearrangement methods

Interval bisection and linear interpolation are fairly straightforward iterative methods for the solution of equations, but neither of them is particularly efficient. You have seen in the examples and exercises that it may easily take six or eight iterations to get a solution accurate to even two decimal places, and in a world where time is at a premium this is not good enough. Other quicker methods must therefore be considered.

Example

Solve the equation $x^5 + 3x^2 - 8 = 0$.

Solution

Let $f(x) = x^5 + 3x^2 - 8$; the graph of $y = f(x)$ shows clearly that there is only one root. Since $f(1) = -4$ and $f(2) = 36$, and since f is a continuous function, the solution lies between 1 and 2, probably close to 1.

Now the equation can be rearranged as $x^5 = 8 - 3x^2$, and this in turn can be written in the form $x = \sqrt[5]{8 - 3x^2}$. This equation can then be used as the basis of an **iteration** formula; i.e. one where, given an approximation x_0, a new approximation x_1 can be calculated, and then a new approximation x_2 can be calculated, etc. In this case the formula is

$$x_{n+1} = \sqrt[5]{8 - 3x_n^2}.$$

Substituting the first approximation $x_0 = 1$ gives $x_1 \approx 1.4$; substituting this result and then each of the others in turn gives $x_2 \approx 1.2$, $x_3 \approx 1.3$, $x_4 \approx 1.24$, $x_5 \approx 1.28$, $x_6 \approx 1.25$, $x_7 \approx 1.27$, $x_8 \approx 1.26$ and $x_9 \approx 1.26$ again. A quick check confirms that $f(1.255) < 0$ and $f(1.265) > 0$, so that the solution is $x \approx 1.26$ correct to two decimal places.

Although this has involved nine calculations (or iterations) and so is apparently no quicker than the previous methods, each iteration involves no more than the substitution of the previous result into a

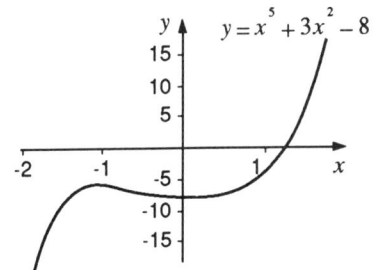

fairly simple formula. It is usually possible, in fact, to make the substitution using the value in the calculator directly, without the trouble of writing down each result and re-entering it: certainly a very simple program can be written for a programmable calculator or computer.

Example

Solve $x^2 + \sin x = 1$, with x in radians.

Solution

The equation can be rearranged as $x^2 = 1 - \sin x$, and from a sketch graph this has solutions at $x \approx 0.6$ and at $x \approx -1.4$. The rearrangement leads to an iteration formula

$$x_{n+1} = \sqrt{1 - \sin x_n}$$

and substituting $x_0 = 0.6$ gives in turn $x_1 \approx 0.66$, $x_2 \approx 0.622$, $x_3 \approx 0.646$, $x_4 \approx 0.631$, $x_5 \approx 0.640$, and $x_6 \approx 0.635$. Once again it is now easy to check that 0.635 and 0.645 give $x^2 + \sin x$ values respectively less than and greater than 1, so that the positive solution is $x \approx 0.64$ to two decimal places.

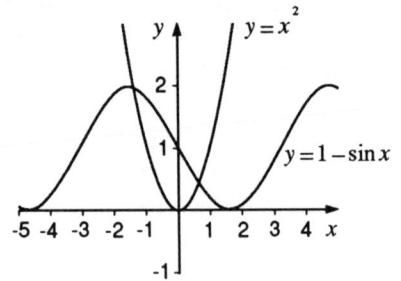

Substituting $x_0 = -1.4$, on the other hand, gives $x_1 \approx 1.41$ and $x_2 \approx 0.11$ and gradually moves in to the same solution as before. This is because the iteration formula has taken the positive square root, and so naturally gives only positive results. An alternative iteration formula with a negative square root would be equally valid:

$$x_{n+1} = -\sqrt{1 - \sin x_n}$$

and with $x_0 = -1.4$ this gives $x_1 \approx x_2 \approx -1.41$, which can be checked as correct in the usual way.

Exercise 3E

1. Show that the equation $x^3 + 2x - 6 = 0$ has just one solution, and locate it. Show that the equation can lead to the iteration formula
$$x_{n+1} = \sqrt[3]{6 - 2x_n},$$
and use this formula to find the solution correct to two decimal places.

2. Write down the equation whose solution can be found by using the iteration formula
$$x_{n+1} = 1 + \frac{1}{x_n^2},$$
and use the formula to find this solution correct to two decimal places.

3. Show that the equation $x^4 - 3x + 1 = 0$ has two solutions, and locate them approximately. Show that the equation can lead to two iteration formulae
$$x_{n+1} = \sqrt[4]{3x_n - 1} \quad \text{and} \quad x_{n+1} = \frac{x_n^4 + 1}{3}$$
and show that each of these formulae leads to just one solution. Find each solution correct to two decimal places.

4. Find a suitable iteration formula for the equation $x^3 - x^2 - 5 = 0$, and solve the equation correct to two decimal places.

5. Use an iterative method to solve the equation $x^x = 2$ correct to three decimal places.

*3.7 Convergence

Exercise 3E should have started you thinking. Why is it, for example, that in Question 3 one iteration formula would work only for the smaller root and the other only for the greater? Why do some formulae seem to work faster than others? Why do some formulae not lead to a root at all, but give results that swing wildly from side to side?

These are all questions concerned with the **convergence** of the iterative process, and while it is useful for you to have a general understanding of the idea, you do not need to go deeply into the theory. If you want to study convergence more deeply than this section allows, look at an undergraduate textbook on numerical analysis.

Consider the equation $x^2 - 5x + 2 = 0$, to be solved using the formula $x_{n+1} = \sqrt{5x_n - 2}$. The diagram shows two graphs, those of $y = x$ and $y = \sqrt{5x - 2}$ respectively; their points of intersection correspond to the solutions of the equation. The first approximation x_0 is substituted into the formula and gives a value - call it y_0 - which then becomes x_1 : this process is illustrated by the arrows.

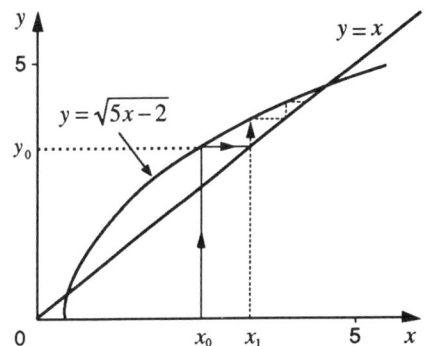

The principle of iteration

As the iteration is repeated, the arrows build into some sort of pattern. In this particular case, they are moving (slowly) closer to the right-hand point of intersection, and so will eventually lead to that solution.

If you try to apply the same method to find the left-hand point of intersection, you will fail no matter how hard you try - the arrows will either converge onto the right-hand point or diverge altogether. On the other hand, the iteration formula

$$x_{n+1} = \frac{x_n^2 + 2}{5}$$

(illustrated in a similar way in the second diagram) converges quite neatly onto the left-hand point of intersection and cannot be persuaded to lead to the upper solution.

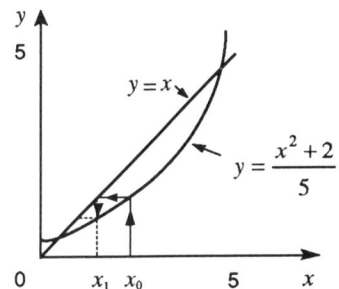

An alternative iteration formula for the lower root

Test for convergence

If you want to know whether or not a particular iteration formula will converge, it is often easiest in practice to apply the formula three or four times and look at the pattern of results. If the results are getting gradually closer together they are probably converging on a solution; if not, you are unlikely to find the solution using this formula. There is a more formal test, however, that you might want to use in a particular case.

Suppose that the iteration formula is $x_{n+1} = g(x_n)$, and that the derivative of $g(x)$ is $g'(x)$. Then it can be shown - the proof is not attempted here - that a **necessary** and **sufficient** condition for the formula to **converge** on the true solution λ is that

$$-1 < g'(\lambda) < 1.$$

This condition is all very well, except that the value of λ is what you are trying to find! In practice, therefore, it is usual to look for an x-interval containing both the unknown λ and the first approximation x_0 such that $|g'(x)| < 1$ throughout the interval. In the example above, if $g(x) = \sqrt{5x - 2}$ then $g'(x) = \dfrac{5}{2\sqrt{5x - 2}}$.

This has absolute value less than 1 if $x > 1.65$, and so can converge on the upper solution (given a suitable first approximation) but not the lower. On the other hand, if $g(x) = \dfrac{x^2 + 2}{5}$ then $g'(x) = \dfrac{2x}{5}$, which lies between −1 and 1 only when $-2.5 < x < 2.5$; it thus converges only to the lower solution.

*Exercise 3F

Use a graph to locate approximately the root or roots of each of the equations opposite. Rearrange each equation to give an iteration formula, and test each formula to determine whether it will lead to any or all of the roots. Repeat with another rearrangement if necessary, until all the roots are obtainable. Find each root correct to one decimal place.

1. $x^3 - 3x - 4 = 0$
2. $x^4 + 2x^3 = 5$.
3. $3^x = 3x + 2$.

3.8   Newton's method

The diagram shows part of the curve $y = f(x)$, where the equation to be solved is $f(x) = 0$. It also shows an approximate solution x_n and the tangent to the curve at the point (x_n, y_n). If this tangent cuts the x-axis at the point $(x_{n+1}, 0)$, then it is clear in this case that x_{n+1} is a better approximation than x_n to the true root λ.

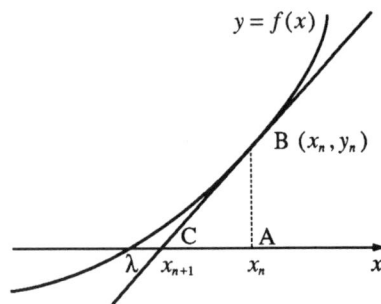

Newton's method for solving $f(x) = 0$

Now the gradient of BC is $\dfrac{BA}{AC}$, which is $\dfrac{y_n}{(x_n - x_{n+1})}$.

But $y_n = f(x_n)$, and the gradient of the tangent is $f'(x_n)$,

so $f'(x_n) = \dfrac{f(x_n)}{(x_n - x_{n+1})}$.

Rearranging this gives **Newton's iteration** formula (sometimes known as the Newton-Raphson formula):

$$x_{n+1} = x_n - \frac{f(x_n)}{f'(x_n)}$$

This can be used to obtain a sequence of results leading to a root in the same way as the iteration formulae discussed in Section 3.6; its main advantage over those formulae is that it tends to converge much more quickly.

Example

Solve the equation $x^3 + 3x^2 - 12 = 0$, correct to two decimal places.

Solution

From a graph, there is just one solution, which lies between 1 and 2. Take $x_0 = 1.5$ as a first approximation.

$f(x) = x^3 + 3x^2 - 12$, so $f'(x) = 3x^2 + 6x$.

Applying Newton's formula,

$$\begin{aligned}
x_1 &= 1.5 - \frac{f(1.5)}{f'(1.5)} \\
&= 1.5 - \frac{-1.875}{15.75} \\
&= 1.62.
\end{aligned}$$

Applying the formula again,

$$\begin{aligned}
x_2 &= 1.62 - \frac{f(1.62)}{f'(1.62)} \\
&= 1.62 - \frac{0.125}{17.59} \\
&= 1.613.
\end{aligned}$$

And it is not difficult now, after just two iterations, to check that $f(1.605) < 0$ and that $f(1.615) > 0$, so giving $x \approx 1.61$ correct to two decimal places.

Example

Solve the equation $x^5 = 5x^2 - 2$, giving your answers correct to three decimal places.

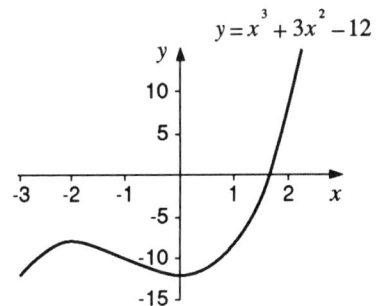

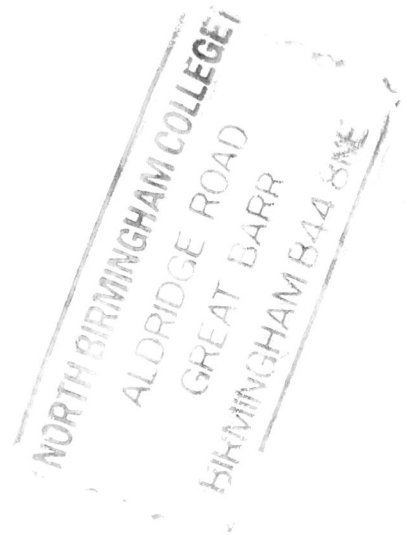

Solution

From a graph, there are solutions at $x \approx -0.6$, $x \approx 0.6$ and $x \approx 1.6$ respectively. Newton's method works only for equations in the form $f(x) = 0$, so a rearrangement gives $f(x) = x^5 - 5x^2 + 2$ and $f'(x) = 5x^4 - 10x$.

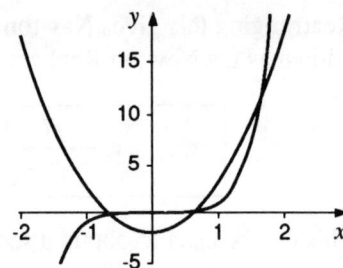

If $x_0 = -0.6$, then

$$x_1 = -0.6 - \frac{f(-0.6)}{f'(-0.6)}$$

$$= -0.6 - \frac{0.122}{6.648}$$

$$= -0.618.$$

If $x_1 = -0.618$, then

$$x_2 = -0.618 - \frac{f(-0.618)}{f'(-0.186)}$$

$$= -0.618 - \frac{0.00023}{6.909}$$

$$= -0.618.$$

$f'(-0.6185) < 0$ and $f'(-06175) > 0$, so $x \approx -0.618$ to three decimal places.

If $x_0 = 0.6$ then $x_1 = x_2 = 0.651$; and if $x_0 = 1.6$ then $x_1 = 1.619$ and $x_2 = 1.618$, which can similarly be confirmed as sufficiently accurate. Thus $x \approx -0.618, 0.651$ or 1.618 to three decimal places.

Example

Solve $\cos x = x^3$, where x is in radians, correct to three decimal places.

Solution

A sketch graph shows just one root, close to 0.9. Rearranging the equation gives $f(x) = \cos x - x^3$ and $f'(x) = -\sin x - 3x^2$.

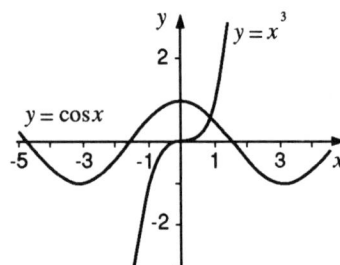

If $x_0 = 0.9$, $x_1 = 0.9 - \dfrac{-0.107}{-3.213} = 0.867$

$x_2 = 0.867 - \dfrac{-0.0046}{-3.017} = 0.866$

$x_3 = 0.866 - \dfrac{-0.0016}{-3.0116} = 0.865.$

Since $f(0.8645) > 0$ and $f(0.8655) < 0$, $x \approx 0.865$ correct to three decimal places.

Evaluation of Newton's method

There is no doubt that Newton's method is a very useful one, and it does have certain advantages over the methods discussed earlier. It converges much faster than any of the previous methods and also the same formula can be used for each of the roots.

The method is not perfect, however, and it does have a number of disadvantages. The first of these is that $f(x)$ has to be differentiated, and your ability to do this will depend on your knowledge of calculus techniques. Then, the first approximation must normally be fairly close to the root you are trying to find, making a reasonable graph almost essential. Finally, the method can become unreliable if the graph of $y = f(x)$ has a turning point or inflexion close to the root - in such a case a different iterative method may prove more effective.

In general, however, you will find the most effective strategy for the solution of difficult equations to be the following:

1. Draw a graph or graphs to locate the root(s) approximately.

2. Use a change of sign and a single linear interpolation to get a good first approximation.

3. Apply Newton's method once or twice (or more).

4. Verify that your answer has the accuracy you require.

*The secant method

A variation of Newton's method which avoids the need for differentiation is the **secant method** - a secant is a chord which continues beyond the curve. If x_n and x_{n+1} are two approximations to a root λ of the equation $f(x) = 0$, the method uses the formula

$$x_{n+2} = x_{n+1} - \frac{(x_n - x_{n+1})f(x_{n+1})}{f(x_n) - f(x_{n+1})}$$

to give a better approximation.

The justification of this formula can be found in the diagram opposite. Using similar triangles, $\dfrac{AB}{BE} = \dfrac{ED}{DF}$,

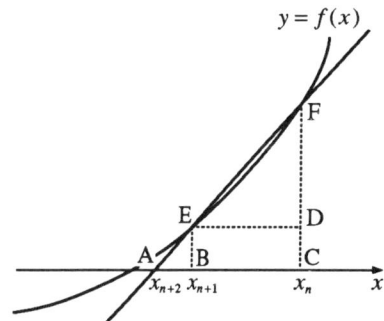

The secant method

therefore
$$\frac{x_{n+1}-x_{n+2}}{f(x_{n+1})} = \frac{x_n - x_{n+1}}{f(x_n)-f(x_{n+1})}$$

from which the formula is easily deduced.

Example

Find a root of the equation $\ln x = \dfrac{1}{1+x^2}$.

Solution

A graph shows only one root, close to 1.4. Although it is possible to differentiate $f(x) \equiv \ln x - \dfrac{1}{1+x^2}$, this is not particularly easy and the secant method can be used instead.

Taking $x_0 = 1.5$ and $x_1 = 1.4$,

$$x_2 = 1.4 - \frac{(1.5-1.4)f(1.4)}{f(1.5)-f(1.4)}$$

$$= 1.4 - \frac{0.1 \times -0.00137}{0.09777-(-0.00137)}$$

$$= 1.4014.$$

$$x_3 = 1.4014 - \frac{(1.4-1.4014)f(1.4014)}{f(1.4)-f(1.4014)}$$

$$= 1.4014 - \frac{-0.0014 \times 0.0000809}{-0.00137-0.0000809}$$

$$= 1.4013.$$

You can check in the usual way that $x \approx 1.4013$ is a solution correct to four decimal places.

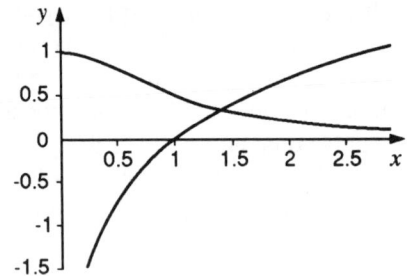

Exercise 3G

1. Show that the equation $x^3 - 2x^2 + 4 = 0$ has a root close to $x = -1$. Use Newton's method to find this root correct to two decimal places.

2. Find correct to three decimal places the smallest positive root of the equation $x^4 - 3x^3 + 5x^2 - 1 = 0$.

*3. Use Newton's method to find both solutions of the equation $x = 3\ln x$ to three decimal places.

4. Find $\sqrt[3]{10}$ using $x_0 = 2$ and two applications of Newton's method, and calculate the percentage error from the 'true' value obtained from a calculator.

*5. Solve $x^x = 5$ correct to four decimal places.

3.9   The ladders again

With the techniques covered in this chapter it is possible to complete the solution of the ladder problem introduced in Section 3.1. The problem, you will recall, was as follows:

In a narrow passage between two walls there are two wooden ladders, a green one 3 m long and a red one 2 m long. Each ladder has its foot at the bottom of one wall and its top resting against the other wall. The green ladder slopes up from left to right and the red ladder slopes up from right to left. The ladders cross 1 m above the ground. How wide is the passage?

This problem led to the equation

$$\frac{1}{\sqrt{4-x^2}} + \frac{1}{\sqrt{9-x^2}} = 1$$

which was left unsolved at that time.

Although it is possible to attempt a graphical or numerical solution of the equation as it stands, it is probably better to simplify it by getting rid of the fractions and the roots. Multiplying the whole equation by both square roots,

$$\sqrt{9-x^2} + \sqrt{4-x^2} = \sqrt{4-x^2}\,\sqrt{9-x^2}.$$

Squaring,

$$\left(9-x^2\right) + 2\sqrt{9-x^2}\,\sqrt{4-x^2} + \left(4-x^2\right) = \left(4-x^2\right)\left(9-x^2\right).$$

Collecting terms,

$$2\sqrt{9-x^2}\,\sqrt{4-x^2} = \left(4-x^2\right)\left(9-x^2\right) - \left(9-x^2\right) - \left(4-x^2\right)$$

$$= 23 - 11x^2 + x^4.$$

Squaring again,

$$4\left(9-x^2\right)\left(4-x^2\right) = \left(23 - 11x^2 + x^4\right)^2$$

$$144 - 52x^2 + 4x^4 = 529 - 506x^2 + 167x^4 - 22x^6 + x^8$$

$$\therefore \quad x^8 - 22x^6 + 163x^4 - 454x^2 + 385 = 0.$$

Now let $f(x) = x^8 - 22x^6 + 163x^4 - 454x^2 + 385$ and draw the graph of $y = f(x)$ for $0 \le x \le 2$, since any valid solution must certainly lie within these bounds.

The diagram shows the result, and it is clear that there are two solutions, close to $x = 1.2$ and $x = 1.9$ respectively.

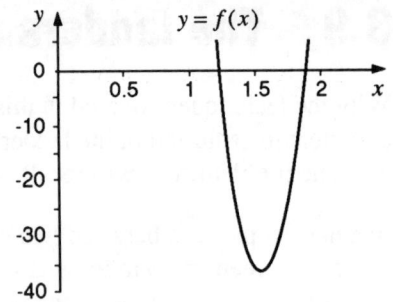

Graph of function for $0 \le x \le 2$

Now $f'(x) = 8x^7 - 132x^5 + 652x^3 - 908x$: this is the step which perhaps justifies the algebra, because differentiating the original equation would have been very messy.

Applying the Newton-Raphson formula with $x_0 = 1.2$,

$$x = 1.2 - \frac{f(1.2)}{f'(1.2)}$$

$$= 1.2 - \frac{7.84}{-263}$$

$$= 1.23 \text{ to two decimal places.}$$

Similarly,

$$x_2 = 1.23 - \frac{f(1.23)}{f'(1.23)}$$

$$= 1.23 - \frac{0.285}{-241}$$

$$= 1.23 \text{ again.}$$

It is easy to check that $f(1.225) > 0$ and $f(1.235) < 0$, confirming that this solution is accurate to the nearest centimetre.

Similarly, taking $x_0 = 1.9$,

$$x_1 = 1.9 - \frac{f(1.9)}{f'(1.9)}$$

$$= 1.9 - \frac{5.12}{194}$$

$$= 1.87 \text{ to two decimal places.}$$

A second iteration gives 1.87 again and this solution of $f(x) = 0$ can similarly be shown to have the necessary accuracy. However, a few moments of thought, possibly by sketching the positions of the ladders, will show you that the passage certainly cannot be 1.87 m wide. Sometimes squaring the original equation (as we have done in this case) leads to some phantom roots being introduced and so one ought to spend a little time checking that the answer obtained does solve the original problems.

The other root of $x \approx 1.23$ does work and so the passageway must be approximately 123 cm wide.

*Activity 4 Iterative chaos

Use a computer or a programmable calculator to investigate sequences given by the iteration formula $x_{n+1} = kx_n(1-x_n)$, with $x_0 = 0.7$, for different values of k between 1 and 4. The results could be quite chaotic!

3.10 Miscellaneous Exercises

1. Show that the equation $x^3 - x - 2 = 0$ has only one real root, and that this root lies between 1 and 2. Use an iterative method to determine its value accurate to three decimal places.

2. Determine graphically the number of solutions of the equation
$$x^2 - 4 = \frac{1}{x},$$
and estimate their values.

3. Show that the equation $2^x = 3x + 2$ has two solutions, and find each of them correct to two decimal places.

4. Find correct to two decimal places all the solutions of the equation
$$x^4 + 2x^3 - 11x^2 - 12x + 21 = 0.$$

5. Find correct to four decimal places the smallest positive root of the equation
$$7x^3 - 19x^2 + 14x - 3 = 0.$$

*6. Solve the equation $xe^x = 1$, correct to two decimal places.

7. Without using any calculator functions other than +, −, × and ÷, find $\sqrt{3}$ correct to six decimal places.

8. Find correct to two decimal places the coordinates of the points at which the circle $x^2 + y^2 = 16$ meets the rectangular hyperbola $x(y+1) = 9$.

*9. Janet and John each have £100 to invest. Janet puts her money in a savings account paying 5% per annum compound interest, while John goes for an account paying 10% per annum simple interest. When (other than at the very beginning) will their respective accounts contain equal balances?

*10. In a circle whose radius is 10 cm, a segment of area 50 cm² is cut off by a chord AB. Show that AB subtends an angle θ radians at the centre of the circle, where $\theta - \sin\theta = 1$. Solve this equation and hence find the perimeter of the segment in cm correct to two decimal places.

4 INEQUALITIES

Objectives

After studying this chapter you should

- be able to manipulate simple inequalities;
- be able to identify regions defined by inequality constraints;
- be able to use arithmetic and geometric means;
- be able to use inequalities in problem solving.

4.0   Introduction

Since the origin of mankind the concept of one quantity being greater than, equal to or less than, another must have been present. Human greed and 'survival of the fittest' imply an understanding of inequality, and even as long ago as 250 BC, Archimedes was able to state the inequality

$$3\frac{10}{71} < \pi < 3\frac{10}{70}.$$

Nowadays we tend to take inequalities for granted, but the concept of **inequality** is just as fundamental as that of **equality**.

You certainly meet inequalities throughout life, though often without too much thought. For example, in the United Kingdom the temperature T°C is usually in the range

$$-15 < T < 30$$

and it would be extremely cold or hot if the temperature was outside this range. In fact, animal life can exist only in the narrow band of temperature defined by

$$-60 < T < 60.$$

Only two other planets in our solar system, Mars and Venus, have temperatures which overlap this band.

It will be assumed that you are familiar with a basic understanding of the use of inequalities $<$, $\leq$, $>$ and $\geq$, and that you have already met the graphical illustration of simple inequalities. You will cover this ground again, but experience with this using Cartesian coordinates and some competence in algebraic manipulation would be very helpful.

Activity 1

Find and prove an inequality relationship for π.

4.1 Fundamentals

The concept of 'greater than' or 'less than' enables numbers to be ordered, and represented on, for example, a number line.

The time line opposite gives a time scale for some important events.

You can also use inequalities for other quantities. For example, the speed of a small car will normally lie within the limits

$$-15\text{mph} \ < \ \text{speed} \ < \ 120\text{mph}.$$

Before looking at more inequality relationships the definition must be clarified. Writing $x > y$ simply means that $x - y$ is a positive number: the other inequalities $<$, $\geq$, $\leq$ can be defined in a similar way. Using this definition, together with the fact that the sum, product and quotient of two positive numbers are all positive, you can prove various inequality relationships.

Example

Show that

(a) if $u > v$ and $x > y$ then $u + x > v + y$;

(b) if $x > y$ and k is a positive number, then $kx > ky$.

Solution

(a) If $v > v$ and $x > y$ then this simply means that $u - v$ and $x - y$ are both positive numbers: hence their sum

$$u - v \text{ and } x - y$$

is also positive. But this can be rewritten as

$$(u + x) - (v + y).$$

Since this difference is a positive number you can deduce that

$$u + x > v + y\text{s}$$

as required.

(b) If $x > y$ then this simply means that $x - y$ is a positive number. Since k is also positive you can deduce that the product $k(x - y)$ is positive. Therefore

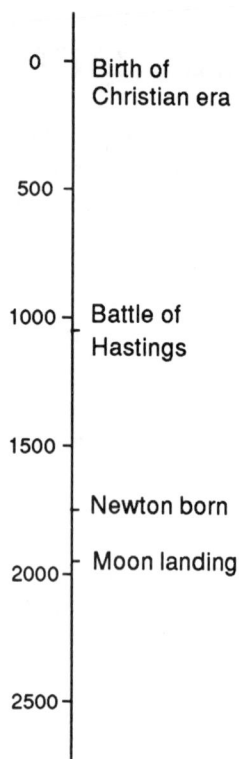

0	Birth of Christian era
500	
1000	Battle of Hastings
1500	
	Newton born
2000	Moon landing
2500	

$$kx - ky = k(x - y)$$

is a positive number, which means that $kx > ky$ as required.

Activity 2

What happens to property (b) when k is a **negative** number?

What happens to subtraction of inequalities? For example, if $u > v$ and $x > y$ then is it always true that $u - v > v - y$?

Can you take square roots through an inequality? i.e. If $a^2 > b^2$ then is it necessarily true that $a > b$?

Investigate these questions with simple illustrations.

In what follows you will need to solve and interpret inequalities. These inequalities will usually be **linear** (that means not involving powers of x, etc), but you will first see how to solve more complex inequalities. The procedure is illustrated in the following example.

Example

Find the values of x which satisfy the inequality

$$x^2 + 7 < 3x + 5.$$

Solution

You can rewrite the inequality as

$$x^2 + 7 - (3x + 5) < 0$$

$$x^2 - 3x + 2 < 0$$

$$(x - 2)(x - 1) < 0.$$

Since the complete expression is required to be negative, this means that one bracket must be positive and the other negative.

This will be the case when $1 < x < 2$.

Exercise 4A

1. Prove that if $x > y > 0$, then $\dfrac{1}{y} > \dfrac{1}{x}$.

2. Prove that if $a^2 > b^2$, where a and b are positive numbers, then $a > b$.

3. Find the values of x for which $8 - x \geq 5x - 4$.

4. Find in each case the set of real values of x for which

   (a) $3(x - 1) \geq x + 1$

   (b) $\dfrac{3}{(x - 1)} \geq \dfrac{1}{(x + 1)}$

5. Find the set of values of x for which
   $$x^2 - 5x + 6 \geq 2.$$

4.2   Graphs of inequalities

In the last section it was shown that inequalities can be solved algebraically; however, it is often more instructive to use a graphical approach.

Consider the previous example in which you want to find values of x which satisfy

$$x^2 + 7 < 3x + 5.$$

Another approach is to draw the graphs of

$$y_1 = x^2 + 7, \quad y_2 = 3x + 5$$

and note when $y_1 < y_2$. This is illustrated in the graph opposite.

Between the points of intersection, A and B, $y_2 > y_1$. Solving the equation $y_1 = y_2$ gives

$$x^2 + 7 = 3x + 5$$

$$\Rightarrow \quad x^2 - 3x + 2 = 0$$

$$\Rightarrow \quad (x - 2)(x - 1) = 0$$

$$\Rightarrow \quad x = 1 \text{ or } 2$$

giving, as before, the solution $1 < x < 2$.

You will find a graphical approach particularly helpful when dealing with inequalities in two variables.

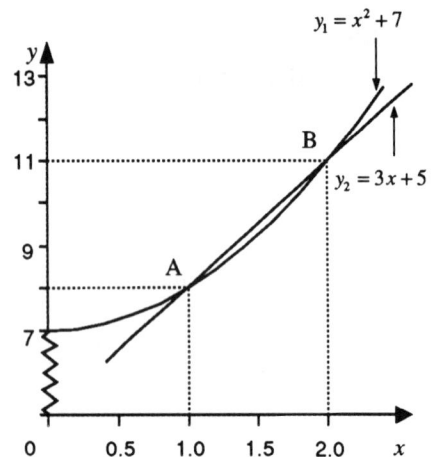

Example

Find the region which satisfies $2x + y > 1$.

Solution

The boundary of the required region is found by solving the **equality**

$$2x + y = 1.$$

This is shown in the diagram opposite

The inequality will be satisfied by all points on one side of the line. To identify which side, you can test the point (0, 0) - this does not satisfy the inequality, so the region to the right of the line is the solution. The **excluded** region is on the **shaded** side of the line.

Just as you can solve simultaneous equations, you can tackle simultaneous inequalities. For example, suppose you require values of x and y which satisfy

$$2x + y > 1$$

and $\qquad x + 2y > 1.$

You have already solved the first inequality, and if you add on the graph of the second inequality, you obtain the region as shown in this diagram.

Combining the two inequalities gives the solution region as shown opposite.

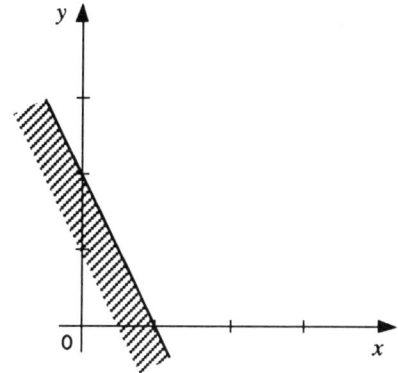

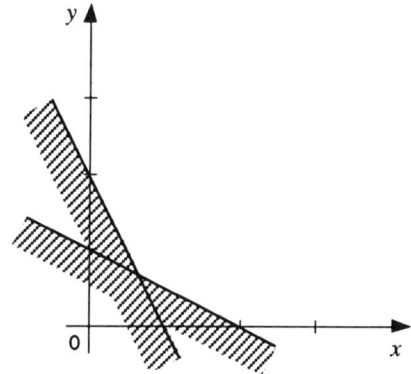

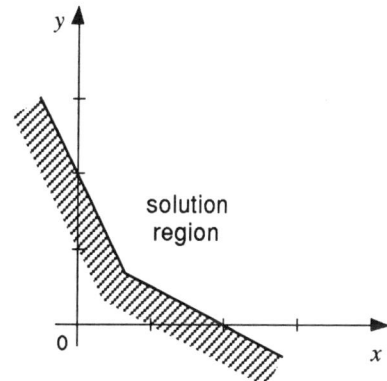

Example

Find the region which satisfies all of the following inequalities.

$$x + y > 2$$
$$3x + y > 3$$
$$x + 3y > 3$$

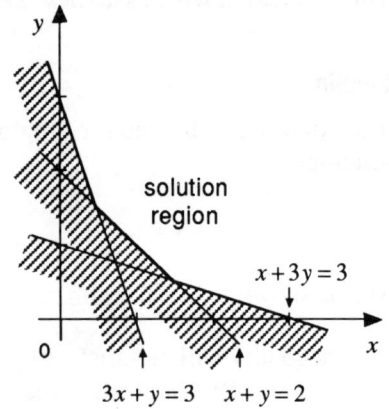

Solution

As before, the graph of the three inequalities is first drawn and the region in which **all** three are satisfied is noted.

Note that if you had wanted to solve

$$x + y < 2$$
$$3x + y > 3$$
$$x + 3y > 3$$

then the solution would have been the triangular region completely bounded by the three lines; in general the word **finite** will be used for such bounded regions.

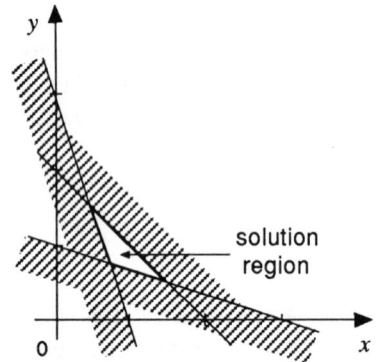

Activity 3

Write down **three** different linear equations of the form

$$ax + by = c.$$

Which three inequalities are satisfied in the finite region formed by these lines?

Suppose you now have **four** linear inequalities in x and y to be satisfied.

What regions might they define?

The following example illustrates some of the possibilities.

Example

In each case find the solution region.

(a) $x + y > 1,\ y - x < 1,\ 2y - x > 0,\ 2x + 3y < 6$

(b) $x + y > 1,\ y - x < 1,\ 2y - x < 0,\ 2x + 3y > 6$

(c) $x + y < 1,\ y - x > 1,\ 2y - x < 0,\ 2x + 3y > 6$

Solution

First graph $x + y = 1$, $y - x = 1$, $2y - x = 0$ and $2x + 3y = 6$, and then in each case identify the appropriate region.

There is no region which satisfies (c).

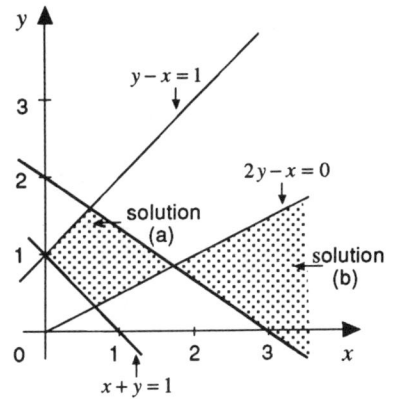

How many finite regions are formed by the intersection of four lines?

Exercise 4B

1. Solve graphically the inequalities

   (a) $(2 - 3x)(1 + x) < 0$

   (b) $x^2 \leq 2x + 8$.

2. Solve

   $y \geq 0$, $x + y \leq 2$ and $y - 2x < 2$.

3. Find the solution set for

   $x + y < 1$ and $3x + 2y < 6$.

4. Find the region which satisfies

   $x + y \geq 2$
   $x + 4y \leq 4$
   $y > -1$.

5. Is the region satisfying
   $x + y > 1$, $3x + 2y < 12$, $y - x < 2$, $2y - x > 1$
   finite?

4.3   Classical inequalities

You are probably familiar with the arithmetic mean (often called the average) of a set of positive numbers. The **arithmetic mean** is defined for positive numbers $x_1, x_2, ..., x_n$ by

$$A = \frac{x_1 + x_2 + ... + x_n}{n}$$

So, for example, if $x_1 = 5$, $x_2 = 6$, $x_3 = 10$, then

$$A = \frac{(5 + 6 + 10)}{3} = 7.$$

There are many other ways of defining a mean; for example, the **geometric mean** is defined as

$$G = \left(x_1 x_2 ... x_n \right)^{1/n}$$

For the previous example,

$$G = (5 \times 6 \times 10)^{\frac{1}{3}} = (300)^{\frac{1}{3}} \approx 6.69$$

The **harmonic mean** is defined by

$$\boxed{\frac{1}{H} = \frac{1}{n}\left(\frac{1}{x_1} + \frac{1}{x_2} + \; \dots \; + \frac{1}{x_n}\right)}$$

So, again with $x_1 = 5$, $x_2 = 6$, and $x_3 = 10$,

$$\frac{1}{H} = \frac{1}{3}\left(\frac{1}{5} + \frac{1}{6} + \frac{1}{10}\right) = \frac{1}{3} \times \frac{7}{15}$$

giving

$$H = \frac{45}{7} \approx 6.43$$

Activity 4

For varying positive numbers x_1, x_2, x_3, find the arithmetic, geometric and harmonic means. What inequality can you conjecture which relates to these three means?

If you have tried a variety of data in Activity 4, you will have realised that the geometric and harmonic means give less emphasis to more extreme numbers. For example, given the numbers 1, 5 and 9,

$$A = 5, \; G = 3.56, \; H = 2.29,$$

whereas for numbers 1, 5 and 102,

$$A = 36, \; G = 7.99, \; H = 2.48.$$

Whilst the arithmetic mean has changed from 5 to 36, the geometric mean has only doubled, and the harmonic mean has hardly changed at all!

In most calculations for mean values the **arithmetic mean** is used, but not always.

One criterion which any mean must satisfy is that, when all the numbers are equal, i.e. when $x_1 = x_2 = \; \dots \; = x_n (= a)$ say, then the mean must equal a.

For example,

$$A = \frac{a+a+\ ...\ +a}{n} = \frac{na}{n} = a$$

$$G = (a\,a\,a\ ...\ a)^{\frac{1}{n}} = \left(a^n\right)^{\frac{1}{n}} = a.$$

Similarly, $H = a$ when all the numbers are equal.

Activity 5

Define a new mean of n positive numbers $x_1, x_2, ..., x_n$ and investigate its properties.

In Activity 4 you might have realised that

$$A \geq G \geq H$$

(equality only occurring when all the numbers are equal). The first inequality will be proved for any two positive numbers, x_1 and x_2.

Given the inequality
$$(x_1 - x_2)^2 \geq 0$$
then equality can occur only when $x_1 = x_2$.

This inequality can be rewritten as

$$x_1^2 - 2x_1x_2 + x_2^2 \geq 0$$

or $\quad x_1^2 + 2x_1x_2 + x_2^2 \geq 4x_1x_2$ (adding $4x_1x_2$ to each side)

giving $\quad \dfrac{(x_1 + x_2)^2}{4} \geq x_1x_2.$

Taking the positive square root of both sides, which was justified in Question 2 of Exercise 4A,

$$\boxed{\frac{x_1 + x_2}{2} \geq \sqrt{x_1x_2}}$$

i.e. $\quad A \geq G$

and equality only occurs when $x_1 = x_2$.

You will see how this result can be used in geometrical problems.

Example

Show that of all rectangles having a given perimeter, the square encloses the greatest area.

Solution

For a rectangle of sides a and b, the perimeter, L is given by

$$L = 2(a+b),$$

and the area, A, by

$$A = ab.$$

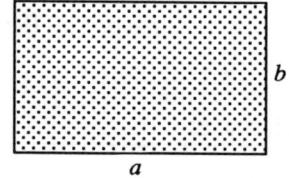

Using the result above, with x_1 replaced by a and x_2 replaced by b,

$$\frac{L}{4} \geq \sqrt{A}$$

or $\quad A \leq \dfrac{L^2}{16}$

where equality occurs only when $a = b$. Since in this example the perimeter is fixed, the right hand side of this last inequality is constant: also equality holds if and only if $a = b$. Therefore you can deduce that the maximum value of A is $\dfrac{L^2}{16}$ and that it is only obtained for the square.

In fact, the inequalities

$$A \geq G \geq H$$

hold for any set of positive numbers, $x_1, x_2, \ldots, x_n$, but the result is not easy to prove, and requires, for example, the use of induction.

The result in the example above illustrates what is called an **isoperimetric inequality**; you will see more of these in the next section.

Exercise 4C

1. Find the arithmetic, geometric and harmonic means for the following sets of numbers, and check that the inequality $A \geq G \geq H$ holds in each case.

   (a) 1, 2, 3, 4;

   (b) 0.1, 2, 3, 4.9;

   (c) 0.1, 2, 3, 100;

   (d) 0.001, 2, 3, 1000;

   (e) 0.001, 0.002, 1000, 2000.

2. Prove that $G \geq H$ for any two positive numbers x_1 and x_2.

*3. By taking functions $\dfrac{1}{x^2}$ and x^2 as numbers in the arithmetic/geometric mean inequality, find the least value of

$$y = \frac{1+x^4}{x^2}.$$

*4. Show that the surface area, S, of a closed cylinder of volume V can be written as

$$S = 2\pi r^2 + \frac{2V}{r}.$$

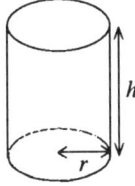

Writing

$$S = 2\pi\left(r^2 + \frac{2V}{2\pi r} + \frac{2V}{2\pi r}\right)$$

and using the arithmetic and geometric means inequality for the three numbers

$$r^2, \frac{V}{2\pi r}, \frac{V}{2\pi r}$$

show that

$$\frac{S}{6\pi} \geq \left(\frac{V^2}{4\pi^2}\right)^{\frac{1}{3}}.$$

When does equality occur? What relationship does this give between h and r?

4.4 Isoperimetric inequalities

In the last section there was an example of an isoperimetric inequality. You will look at a more general result (first known to the Greeks in about 2000 BC) and at some further special cases.

According to legend, Princess Dido was fleeing from the tyranny of her brother and, with her followers, set sail from Greece across the sea. Having arrived at Carthage, she managed to obtain a grudging concession from the local native chief to the effect that

'she could have as much land as could be encompassed by an ox's skin.'

Of course, the natives expected her to kill the biggest ox she could find and use its skin to claim her land - but her followers were very astute, advising her to cut the skin to make as many thin strands as possible and to join them together to form one long length to mark the perimeter of her land. Her only problem then was in deciding what shape this perimeter should be to enclose the maximum area.

What do you think is the best shape?

In mathematical terms, the search is for the shape which maximises the area A inside a given perimeter of length L.

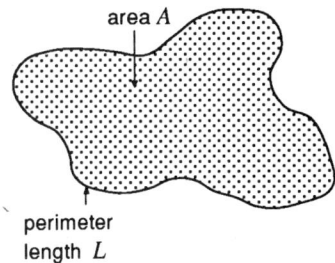

area A

perimeter
length L

Example

For a given perimeter length, say 12 cm, find the area enclosed by

(a) a square;

(b) a circle;

(c) an equilateral triangle.

Solution

(a)   For $L = 12$, each side is of length 3 cm

and $A = 3^2 = 9 \text{ cm}^2.$

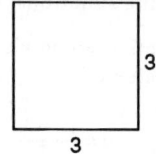

(b)   For $L = 12$, assume the radius is a, giving

$$12 = 2\pi a \Rightarrow a = \frac{6}{\pi}$$

and $A = \pi a^2 = \pi \left(\frac{6}{\pi} \right)^2 = \frac{36}{\pi} \approx 11.46 \text{ cm}^2$

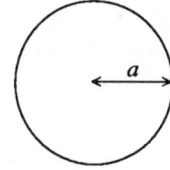

(c)   Again, for $L = 12$, each side is of length 4 cm, and

$$A = \frac{1}{2} \times 4 \times 4 \sin 60 = 4\sqrt{3} \approx 6.93 \text{ cm}^2.$$

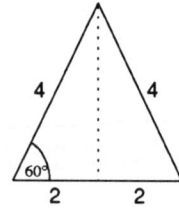

So, for the particular problem of a perimeter length of 12 cm, of the three shapes chosen the circle gives the largest area - but can there be another shape which gives a larger one? You can make some progress by looking more carefully at the circle in the general case of perimeter L.

Now $L = 2\pi a$

and $A = \pi a^2 = \pi \left(\frac{L}{2\pi} \right)^2 = \frac{L^2}{4\pi}$

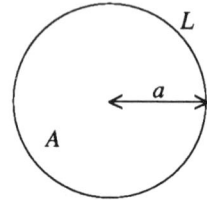

So, for **any** circle

$$\frac{4\pi A}{L^2} = 1.$$

The **Isoperimetric Quotient Number** (I.Q.) of any closed curve is defined as

$$\boxed{\text{I. Q.} = \frac{4\pi A}{L^2}}$$

For the circle, you see that I.Q. = 1. In the basic problem you have been trying to find the shape which gives a maximum value to A for a fixed value of L. In terms of the I.Q. number, you want to find the shape which gives the maximum value to the I.Q. number. But, for a circle, the value of the I.Q. number is 1, so if the optimum shape is a circle, then the inequality

$$\boxed{\text{I.Q.} \leq 1}$$

holds for all plane shapes, and equality occurs **only** for the circle.

Note that, since the I.Q. is the ratio of an area to the square of a length, it is non-dimensional, i.e. a number requiring no units.

Example

Find the I.Q. number for a square of side a.

Solution

$$L = 4a, A = a^2,$$

and

$$\text{I.Q.} = 4\pi \times \frac{a^2}{(4a)^2} = \frac{\pi}{4} \approx 0.785.$$

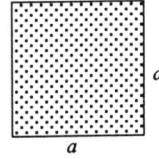

Activity 6

Find I.Q. numbers of various shapes and check that, in each case, the inequality I.Q. ≤ 1 holds.

A complete proof is beyond the scope of this present work (and, in fact, involves high level mathematics). It is surprising that such a simple result, known to the Greeks, could not be proved until the late 19th Century, and even then required sophisticated mathematics. You can, though, verify the result for all regular polygons as will be shown.

Consider a regular polygon of n sides. The angle subtended by each side at the centre is

$$\frac{360}{n} \text{ degrees or } \frac{2\pi}{n} \text{ radians.}$$

You will work in **radians** in what follows. If each side is of length a, the area of each triangle is given by

$$\frac{1}{2} \times a \times \frac{a}{2} \times \frac{1}{\tan\left(\dfrac{\pi}{n}\right)} = \frac{a^2}{4\tan\left(\dfrac{\pi}{n}\right)}$$

The total area, $A = \dfrac{na^2}{4\tan\left(\dfrac{\pi}{n}\right)}$, and $L = na$,

so

$$\text{I.Q.} = 4\pi \times \left(\frac{na^2}{4\tan\left(\dfrac{\pi}{n}\right)}\right) \times \frac{1}{(na)^2}$$

i.e.

$$\text{I.Q.} = \frac{\left(\dfrac{\pi}{n}\right)}{\tan\left(\dfrac{\pi}{n}\right)}$$

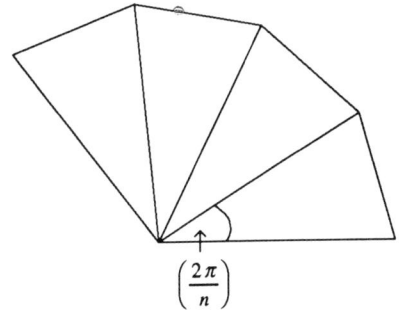

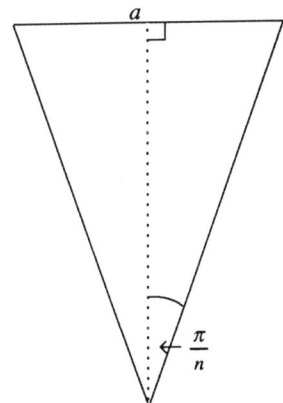

Activity 7

Use your calculator to find the limit of $\left(\dfrac{x}{\tan x}\right)$ as $x \to 0$.

Now you can write $\text{I.Q.} = \dfrac{x}{\tan x}$ where $x = \dfrac{\pi}{n}$.

As $n \to \infty$, the polygon becomes, in the limit, a circle, and you have seen that I.Q.$=1$, as expected. Note that, for all values of the positive integer $n, \tan\left(\dfrac{\pi}{n}\right) > \left(\dfrac{\pi}{n}\right)$ - use your calculator to check some of the values. Hence, for any regular ploygon

$$\text{I.Q.} \leq 1,$$

and you can see that the larger n becomes, the closer the I.Q. comes to 1, fitting in with the fact that the I.Q. for a circle is 1.

Finally, it should be noted that Princess Dido did not live happily ever after. Having been outwitted by her, the native leader promptly fell in love with her. As she did not reciprocate his feelings, she burnt herself on a funeral pyre in order to escape a fate worse than death!

Exercise 4D

1. For a given perimeter length of 12 cm, find the total area enclosed by the rectangle with sides

   (a) 3 cm and 3 cm

   (b) 2 cm and 4 cm

   (c) 1 cm and 5 cm

2. Find the I.Q. numbers for the following shapes:

   (a) equilateral triangle;

   (b) regular hexagon;

   (c) rectangle with sides in the ratio 1 : 2.

3. For a rectangle with sides in the ratio

   $$1 : k \ \ (k \geq 1),$$

   find an expression for the I.Q. number. What value of k gives:

   (a) maximum value

   (b) minimum value

   for the I.Q. number?

4. What is the volume, V, of the sphere which is enclosed by a surface area of 12 cm^2?

5. What is the volume, V, of the cube which is enclosed by a surface area of 12 cm^2?

6. For a given surface area, S, what closed three-dimensional shape do you think gives a maximum volume?

4.5 Miscellaneous Exercises

1. Obtain the sets of values of x for which

   (a) $\quad 2x > \dfrac{1}{x}$

   (b) $\quad \dfrac{1}{x+1} > \dfrac{x}{3+x}.$

2. Find the range of values of x for which

   $4x^2 - 12x + 5 < 0.$

3. Find the ranges of values of x such that

   $x > \dfrac{2}{x-1}.$

4. Find the set of values of x for which

   $\dfrac{x(x+2)}{x-3} < x+1.$

5. Find the solution set of the pair of inequalities

   $x + y < 1$

   $2x + 5y < 10.$

6. Is the region defined by

   $2x - 3y \le 6$

   $x + y \le 4$

   finite?

7. Find the region satisfied by

   $x + y \le 4$

   $2x - 3y \le 6$

   $3x - y \ge -3$

   $x \le 2.$

8. Prove that $A \ge H$ for any two positive numbers. When does equality occur?

9. Find the I.Q. number for the shape illustrated below, where k is a positive constant.

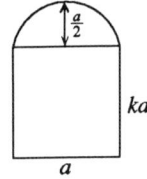

   What value of k gives a maximum I.Q. value?

10. Find the I.Q. number for a variety of triangles, including an equilateral triangle. What do you deduce about the I.Q. numbers for triangles?

*11. For three-dimensional closed shapes, the isosurface area quotient number is defined as

$$\text{I.Q.} = \frac{6\sqrt{\pi}\,V}{S^{\frac{3}{2}}}$$

   where V is the volume enclosed by a total surface area S. Find the I.Q. for a variety of three-dimensional shapes. Can you find an inequality satisfied by all closed shapes in three dimensions?

5 LINEAR PROGRAMMING

Objectives

After studying this chapter you should

* be able to formulate linear programming problems from contextual problems;

* be able to identify feasible regions for linear programming problems;

* be able to find solutions to linear programming problems using graphical means;

* be able to apply the simplex method using slack variables;

* understand the simplex tableau procedure.

5.0 Introduction

The methods of linear programming were originally developed between 1945 and 1955 by American mathematicians to solve problems arising in industry and economic planning. Many such problems involve constraints on the size of the workforce, the quantities of raw materials available, the number of machines available and so on. The problems that will be solved usually have two variables in them and can be solved graphically, but problems occurring in industry have many more variables and have to be solved by computer. For example, in oil refineries, problems arise with hundreds of variables and tens of thousands of constraints.

Another application is in determining the best diet for farm animals such as pigs. In order to maximise the profit a pig farmer needs to ensure that the pigs are fed appropriate food and sufficient quantities of it to produce lean meat. The pigs require a daily allocation of carbohydrate, protein, amino acids, minerals and vitamins. Each involves various components. For example, the mineral content includes calcium, phosphorus, salt, potassium, iron, magnesium, zinc, copper, manganese, iodine, and selenium. All these dietary constituents should be present, in correct amounts.

A statistician at Exeter University has devised a computer program for use by farmers and companies producing animal feeds which enables them to provide the right diet for pigs at various stages of development, such as the weaning, growing and finishing stages. The program involves 20 variables and 10 equations!

Undoubtably linear programming is one of the most widespread methods used to solve management and economic problems, and has been applied in a wide variety of situations and contexts.

5.1 Formation of linear programming problems

You are now in a position to use your knowledge of inequalities from the previous chapter to illustrate **linear programming** with the following case study.

Suppose a manufacturer of printed circuits has a stock of

200 resistors, 120 transistors and 150 capacitors

and is required to produce two types of circuits.

Type A requires 20 resistors, 10 transistors and 10 capacitors.

Type B requires 10 resistors, 20 transistors and 30 capacitors.

If the profit on type A circuits is £5 and that on type B circuits is £12, how many of each circuits should be produced in order to maximise the profit?

You will not actually solve this problem yet, but show how it can be formulated as a linear programming problem. There are three vital stages in the formulation, namely

(a)   What are the unknowns?
(b)   What are the constraints?
(c)   What is the profit/cost to be maximised/minimised?

For this problem,

(a)   **What are the unknowns?**

Clearly the number of type A and type B circuits produced; so we define

x = number of type A circuits produced
y = number of type B circuits produced

(b)   **What are the constraints?**

There are constraints associated with the total number of resistors, transistors and capacitors available.

Resistors   Since each type A requires 20 resistors, and each type B requires 10 resistors, then

$$20x + 10y \le 200,$$

as there is a total of 200 resistors available.

Transistors Similarly

$$10x + 20y \le 120$$

Capacitors Similarly

$$10x + 30y \le 150$$

Finally you must state the obvious (but nevertheless important) inequalities

$$x \ge 0, \ y \ge 0$$

(c) **What is the profit?**

Since each type A gives £5 profit and each type B gives £12 profit, the total profit is £P, where

$$P = 5x + 12y$$

You can now summarise the problem as:

maximise $\quad P = 5x + 12y$

subject to $\quad 20x + 10y \le 200$

$\qquad\qquad 10x + 30y \le 150$

$\qquad\qquad x \ge 0$

$\qquad\qquad y \ge 0.$

This is called a **linear** programming problem since both the objective function P and the constraints are all linear in x and y.

Activity 1   Feasible solutions

Show that $x = 5, y = 3$ satisfies all the constraints.

Find the associated profit for this solution, and compare this profit with other possible solutions.

At this stage, you will not continue with finding the actual solutions but concentrate on further practice in formulating problems of this type.

The key stage is the first one, namely that of identifying the unknowns; so you must carefully read the problem through in order to identify the basic unknowns. Having done this successfully, it should be straightforward to express both the constraints, and profit function in terms of the unknowns.

Example

A small firm builds two types of garden shed.

Type A requires 2 hours of machine time and 5 hours of craftsman time.

Type B requires 3 hours of machine time and 5 hours of craftsman time.

Each day there are 30 hours of machine time available and 60 hours of craftsman time. The profit on each type A shed is £60 and on each type B shed is £84.

Formulate the appropriate linear programming problem.

Solution

(a)   **Unknowns**

Define

x = number of Type A sheds produced each day,

y = number of Type B sheds produced each day.

(b)   **Constraints**

Machine time: $2x + 3y \leq 30$

Craftsman time: $5x + 5y \leq 60$

and $x \geq 0, y \geq 0$

(c)   **Profit**

$P = 60x + 84y$

So, in summary, the linear programming problem is

maximise $P = 60x + 84y$

subject to $2x + 3y \leq 30$

$x + y \leq 12$

$x \geq 0$

$y \geq 0$

Exercise 5A

1. Ann and Margaret run a small business in which they work together making blouses and skirts.

   Each blouse takes 1 hour of Ann's time together with 1 hour of Margaret's time. Each skirt involves Ann for 1 hour and Margaret for half an hour. Ann has 7 hours available each day and Margaret has 5 hours each day.

   They could just make blouses or they could just make skirts or they could make some of each.

   Their first thought was to make the same number of each. But they get £8 profit on a blouse and only £6 on a skirt.

   (a) Formulate the problem as a linear programming problem.

   (b) Find three solutions which satisfy the constraints.

2. A distribution firm has to transport 1200 packages using large vans which can take 200 packages each and small vans which can take 80 packages each. The cost of running each large van is £40 and of each small van is £20. Not more than £300 is to be spent on the job. The number of large vans must not exceed the number of small vans.

   Formulate this problem as a linear programming problem given that the objective is to **minimise** costs.

3. A firm manufactures wood screws and metal screws. All the screws have to pass through a threading machine and a slotting machine. A box of wood screws requires 3 minutes on the slotting machine and 2 minutes on the threading machine. A box of metal screws requires 2 minutes on the slotting machine and 8 minutes on the threading machine. In a week, each machine is available for 60 hours.

   There is a profit of £10 per box on wood screws and £17 per box on metal screws.

   Formulate this problem as a linear programming problem given that the objective is to **maximise** profit.

4. A factory employs unskilled workers earning £135 per week and skilled workers earning £270 per week. It is required to keep the weekly wage bill below £24300.

   The machines require a minimum of 110 operators, of whom at least 40 must be skilled. Union regulations require that the number of skilled workers should be at least half the number of unskilled workers.

   If x is the number of unskilled workers and y the number of skilled workers, write down all the constraints to be satisfied by x and y.

5.2 Graphical solution

In the previous section you worked through problems that led to a linear programming problem in which a **linear** function of x and y is to be maximised (or minimised) subject to a number of **linear** inequalities to be satisfied.

Fortunately problems of this type with just two variables can easily be solved using a graphical method. The method will first be illustrated using the example from the text in Section 5.1. This resulted in the linear programming problem

maximise $P = 5x + 12y$

subject to $20x + 10y \leq 200$

$$10x + 20y \leq 120$$

$$10x + 30y \leq 150$$

$$x \geq 0$$

$$y \geq 0$$

You can illustrate the **feasible** (i.e. allowable) region by graphing all the inequalities and shading out the regions not allowed . This is illustrated in the figure below.

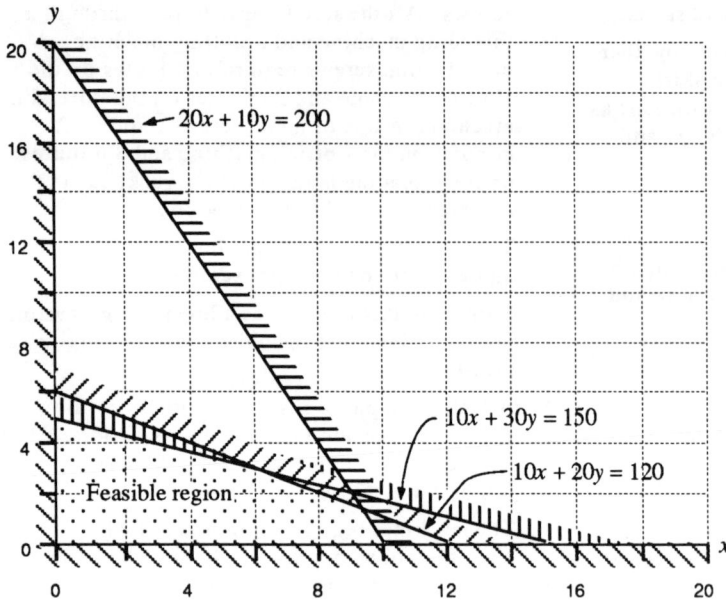

Magnifying the feasible region, you can look at the family of straight lines defined by

$$C = 5x + 12y$$

where C takes various values.

The figure shows, for example, the lines defined by

$$C = 15, \ C = 30, \ \text{and} \ C = 45.$$

On each of these lines any point gives the same profit.

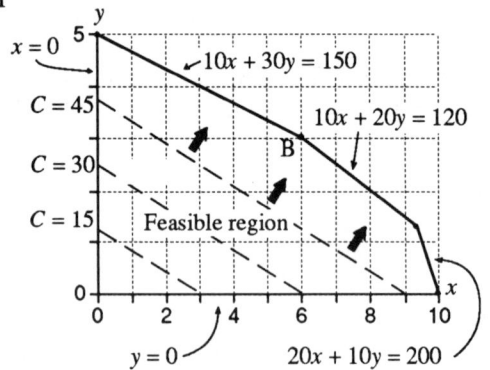

Activity 2

Check that the points

$$x = 1, \ y = \frac{25}{12}$$
$$x = 2, \ y = \frac{5}{3}$$
$$x = 4, \ y = \frac{5}{6}$$

each lie on the curve defined by $C = 30$. What profit does each of these points give?

Where is the point representing maximum profit ?

As the profit line moves to the right, the profit increases and so the maximum profit corresponds to the last point touched as the profit line moves out of the feasible region. This is the point B, the intersection of

$$10x + 30y = 150 \text{ and } 10x + 20y = 120$$

Solving these equations gives $10y = 30$, i.e. $y = 3$ and $x = 6$. So maximum profit occurs at the point $(6, 3)$ and the profit is given by

$$P = 5 \times 6 + 12 \times 3 = 66.$$

Example

A farmer has 20 hectares for growing barley and swedes. He has to decide how much of each to grow. The cost per hectare for barley is £30 and for swedes is £20. The farmer has budgeted £480.

Barley requires 1 man-day per hectare and swedes require 2 man-days per hectare. There are 36 man-days available.

The profit on barley is £100 per hectare and on swedes is £120 per hectare.

Find the number of hectares of each crop the farmer should sow to maximise profits.

Solution

The problem is formulated as a linear programming problem:

(a) **Unknowns**

x = number of hectares of barley

y = number of hectares of swedes

(b) **Constraints**

Land $\quad\quad x + y \le 20$

Cost: $\quad\quad 30x + 20y \le 480$

Manpower $\quad x + 2y \le 36$

(c) **Profit**

$$P = 100x + 120y$$

To summarise, maximise $P = 100x + 120y$

subject to $x + y \le 20$

$30x + 20y \le 480$

$x + 2y \le 36$

$x \ge 0$

$y \ge 0$

The feasible region is identified by the region enclosed by the five inequalities, as shown below. The profit lines are given by

$$C = 100x + 120y$$

and again you can see that C increases as the line (shown dotted) moves to the right. Continuing in this way, the maximum profit will occur at the intersection of

$$x + 2y = 36 \text{ and } x + y = 20$$

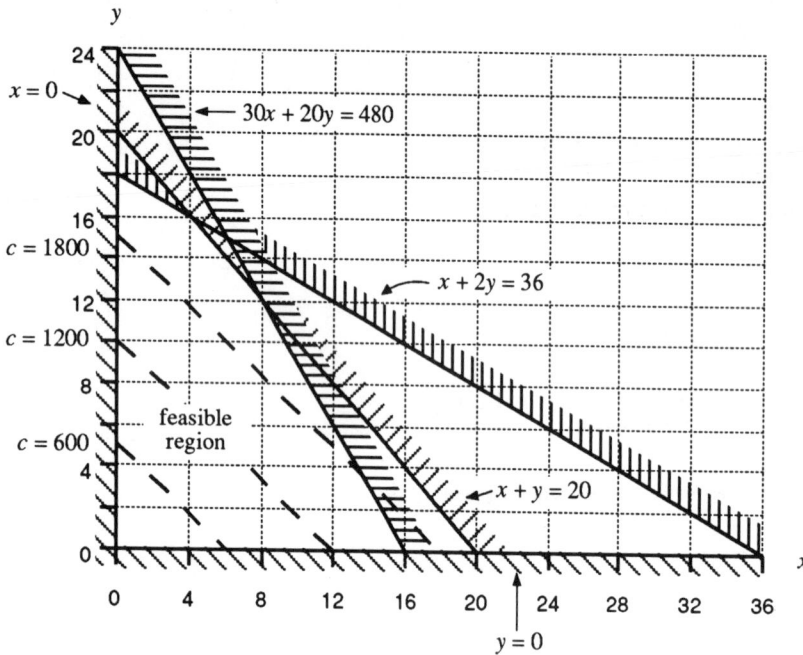

At this point $x = 4$ and $y = 16$, and the corresponding maximum profit is given by

$$P = 100 \times 4 + 120 \times 16 = 2320$$

The farmer should sow 4 hectares with barley and 16 with swedes.

Exercise 5B

1. Solve the linear programming problem defined in Question 1 of Exercise 5A.

2. Solve the linear programming problem defined in Question 2 of Exercise 5A.

3. A camp site for caravans and tents has an area of 1800m² and is subject to the following regulations:

   The number of caravans must not exceed 6.

   Reckoning on 4 persons per caravan and 3 per tent, the total number of persons must not exceed 48.

   At least 200 m² must be available for each caravan and 90 m² for each tent.

   The nightly charges are £2 for a caravan and £1 for a tent.

   Find the greatest possible nightly takings.

   How many caravans and tents should be admitted if the site owner wants to make the maximum profit and have

   (a) as many caravans as possible,

   (b) as many tents as possible?

4. The annual subscription for a tennis club is £20 for adults and £8 for juniors. The club needs to raise at least £800 in subscriptions to cover its expenses.

   The total number of members is restricted to 50. The number of junior members is to be between one quarter and one third of the number of adult members.

   Represent the information graphically and find the numbers of adult and junior members which will bring in the largest amount of money in subscriptions.

   Find also the least total membership which will satisfy the conditions.

5. The numbers of units of vitamins A, B and C in a kilogram of foods X and Y are as follows:

Food	Vitamin A	Vitamin B	Vitamin C
X	5	2	6
Y	4	6	2

   A mixture of the two foods is made which has to contain at least 20 units of vitamin A, at least 24 units of vitamin B and at least 12 units of vitamin C.

   Find the smallest total amount of X and Y to satisfy these constraints.

   Food Y is three times as expensive as Food X. Find the amounts of each to minimise the cost and satisfy the constraints.

5.3 Simplex method

Where will a linear programming solution always occur?

Looking back at the example in Section 5.2, the slope of the profit line was $\frac{5}{6}$. This is more than the slope of the line $x + 2y = 36$ (namely $\frac{1}{2}$), but less than the slope of the other two lines, $x + y = 20$ (i.e 1) and $30x + 20y = 480$ (i.e. $\frac{3}{2}$).

So the solution will occur at the intersection of the two lines with slopes $\frac{1}{2}$ and 1.

Activity 3

Check the slopes of the constraints and profit function in the example in the text in Section 5.2.

Another point worth noting here is that the solution of a linear programming problem will occur at one of the **vertices** of the feasible region.

So an alternative to the graphical method of solution would be to

(a) find all the vertices of the feasible region;

(b) find the value of the profit function at each of these vertices;

(c) choose the one which gives maximum value to the profit function.

You can see how this method works with the second example in Section 5.2. The vertices are given by

(a)   0 (0, 0)

(b)   A (0, 18)

(c)   B (4, 16)

(d)   C (8, 12)

(e)   D (16, 0)

and the corresponding profits in £ are

Point	Profit
0	0
A	2160
B	2320
C	2240
D	1600

As you can see, as you move round the feasible region, the profit increases from 0 to A to B, but then decreases to C to D and back to 0.

In more complicated problems, it is helpful to introduce the idea of **slack variables**. For the problem above, with the three inequalities

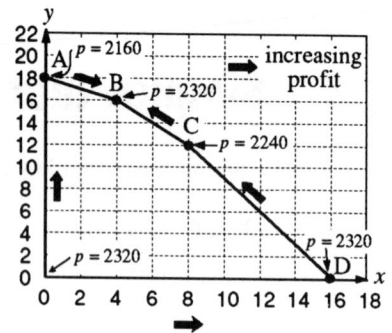

$$x + y \leq 20$$
$$30x + 20y \leq 480$$
$$x + 2y \leq 36$$

three new variables are defined by

$$r = 20 - x - y$$
$$s = 480 - 30x - 20y$$
$$t = 36 - x - 2y$$

The three inequalities can now be written as

$$r \geq 0, \; s \geq 0, \; t \geq 0$$

as well as $x \geq 0$, $y \geq 0$. The variables r, s and t are called the slack variables as they represent the amount of slack between the total quantity available and how much is being used.

The importance of the slack variables is that you can now define each vertex in terms of two of the variables, $x, y, r, s,$ or t, being zero; for example, at A,

$$x = t = 0$$

Activity 4

Complete the table below, defining each vertex

$$0 \quad x = y = 0$$
$$A \quad t = x = 0$$
$$B \quad \dots\dots\dots$$
$$C \quad \dots\dots\dots$$
$$D \quad \dots\dots\dots$$

The procedure of increasing the profit from one vertex to the next will be again followed. Starting at the origin

$$P = 100x + 120y = 0 \text{ at } x = y = 0$$

Clearly P will increase in either direction from the origin - moving up the y axis means that x is held at zero whilst y increases. Although you know from the diagram that the next vertex reached will be A, how could you work that our without a picture? As x is being kept at zero and y is increasing the next vertex met will either be where $x = r = 0$ or where $x = s = 0$ or where $x = t = 0$. But note that

$$x = r = 0 \Rightarrow y = 20$$
$$x = s = 0 \Rightarrow y = 24$$
$$x = t = 0 \Rightarrow y = 18 \leftarrow \text{ smallest}$$

and so as y increases the first of these three points it reaches is the one where y is the smallest, namely $y = 18$ and the vertex is where $x = t = 0$ (which the point A, thus confirming without a picture what has already been seen). You can now express the profit P in terms of x and t:

$$P = 100x + 120y$$
$$= 100x + 120\frac{(36 - x - t)}{2}$$
$$= 40x - 60t + 2160$$

How can you tell if P will increase as x increases?

So you now increase x, keeping t at zero. You will next meet a vertex where either $t = s = 0$, $t = r = 0$ or $t = y = 0$. As x is increasing the vertex will be the one of those three points where x is the smallest:

$$t = s = 0 \Rightarrow x = 6$$
$$t = r = 0 \Rightarrow x = 4 \leftarrow \text{ smallest}$$
$$t = y = 0 \Rightarrow x = 18$$

So the next vertex reached is where $t = r = 0$, (which is B) and P is now expressed in terms of t and r. To do this eliminate y from the first and third of the original equations by noting that

$$2r - t = 2(20 - x - y) - (36 - x - 2y) = 4 - x$$

Hence

$$P = 40x - 60t + 2160$$
$$= 40(4 - 2r + t) - 60t + 2160$$
$$= -80r - 20t + 2320$$

How can you tell that P has reached its maximum?

Throughout the feasible region you know that $r \geq 0$ and $t \geq 0$ and so it is clear from the negative coefficients in the above expression for P that P reaches its maximum value of 2320 when r and t are both zero. This happens when $x = 4$ and $y = 16$.

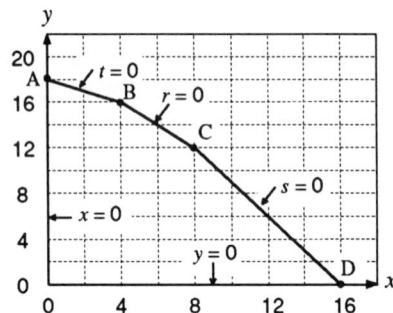

Activity 5

Now travel round the feasible region from 0 to D to C to B. At each vertex, express P in terms of the defining variables, and check that P will continue to increase until B is reached.

Although this probably looks a much more complicated way of solving linear programming problems, its real application is to problems of more than 2 variables. These cannot be solved graphically, but can be solved using a procedure using slack variables called the **simplex** method.

Exercise 5C

1. (a) Solve the linear programming problems
maximise $P = 2x + 4y$

subject to $x + 5y \leq 10$
$$4x + y \leq 8$$
$$x \geq 0$$
$$y \geq 0$$

by a graphical method.

(b) Introduce slack variables r and s, and solve the problem by the simplex method.

2. (a) Determine the vertices of the feasible region for the linear programming problem
maximise $P = x + y$

subject to $x + 4y \leq 8$
$$2x + 3y \leq 12$$
$$3x + y \leq 9$$
$$x \geq 0$$
$$y \geq 0$$

Hence find the solution.

(b) Verify this solution by using the simplex method.

3. Use the simplex method to solve the linear programming problem:

maximise $P = 10x + 15y$

subject to $4y + 10x \leq 40$
$10y + 3x \leq 30$
$5y + 4x \leq 20$
$x \geq 0$
$y \geq 0$

*5.4 Simplex tableau

The way this method works will be illustrated with the example.

Maximise $P = x + 2y$

subject to $x + 4y \leq 20$
$x + y \leq 8$
$5x + y \leq 32$
$x \geq 0$
$y \geq 0$

As usual introduce slack variables r, s and t defined by

$$x + 4y + r = 20$$
$$x + y + s = 8$$
$$5x + y + t = 32$$

and write the equations in the matrix form

$$P - x - 2y = 0$$
$$x + 4y + r = 20$$
$$x + y + s = 8$$
$$5x + y + t = 32$$

$$\Rightarrow \begin{bmatrix} 1 & -1 & -2 & 0 & 0 & 0 \\ 0 & 1 & 4 & 1 & 0 & 0 \\ 0 & 1 & 1 & 0 & 1 & 0 \\ 0 & 5 & 1 & 0 & 0 & 1 \end{bmatrix} \begin{bmatrix} P \\ x \\ y \\ r \\ s \\ t \end{bmatrix} = \begin{bmatrix} 0 \\ 20 \\ 8 \\ 32 \end{bmatrix}$$

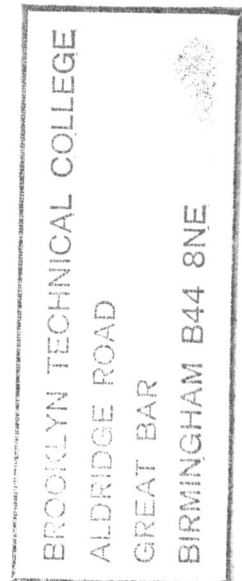

The **augmented matrix** with the extra right hand column will be used.

P	x	y	r	s	t			Comments
1	-1	-2	0	0	0	0		Increase x first (y could have been chosen) and compare values of x where $y=s=0$, and
0	1	4	1	0	0	20		$y=t=0$, namely $20/1, 8/1, 32/5$*. These values are easily spotted from the matrix as the R.H.
0	1	1	0	1	0	8		figures divided by the corresponding coefficients of x.
0	5	1	0	0	1	32		(*This term has the smallest positive value so now manipulate the matrix to express P in terms of y and t)
1	-1	-2	0	0	0	0		
0	1	4	1	0	0	20		
0	1	1	0	1	0	8		
0	1	$\frac{1}{5}$	0	0	$\frac{1}{5}$	$\frac{32}{5}$	$\leftarrow R_4/5$	
1	0	$-\frac{9}{5}$	0	0	$\frac{1}{5}$	$\frac{32}{5}$	$\leftarrow R_1+R_4$	From the first row express P in terms of y and t, with a positive coefficient of y. Increase y and
0	0	$\frac{19}{5}$	1	0	$-\frac{1}{5}$	$\frac{68}{5}$	$\leftarrow R_2-R_4$	compare the values obtained from the figures in the R.H. column divided by the corresponding
0	0	$\frac{4}{5}$	0	1	$-\frac{1}{5}$	$\frac{8}{5}$	$\leftarrow R_3-R_4$	coefficients of y, namely $(68/5)/(19/5)$, $(8/5)/(4/5)$*, $(32/5)/(1/5)$.
0	1	$\frac{1}{5}$	0	0	$\frac{1}{5}$	$\frac{32}{5}$		(*This term has the smallest positive value (where $s=t=0$) so now manipulate the matrix to express P in terms of s and t)
1	0	$-\frac{9}{5}$	0	0	$\frac{1}{5}$	$\frac{32}{5}$		
0	0	$\frac{19}{5}$	1	0	$-\frac{1}{5}$	$\frac{68}{5}$		
0	0	1	0	$\frac{5}{4}$	$-\frac{1}{4}$	2	$\leftarrow R_3/(4/5)$	
0	1	$\frac{1}{5}$	0	0	$\frac{1}{5}$	$\frac{32}{5}$		
1	0	0	0	$\frac{9}{4}$	$-\frac{1}{4}$	10	$\leftarrow R_1+\frac{9}{5}R_3$	From the first row P could now be expressed
0	0	0	1	$-\frac{19}{4}$	$\frac{3}{4}$	6	$\leftarrow R_2-\frac{19}{5}R_3$	in terms of s and t, with a positive coefficient of t. So now increase t and compare the values
0	0	1	0	$\frac{5}{4}$	$-\frac{1}{4}$	2		$6/(3/4)$*, $2/(-1/4)$, $6/(1/4)$ (*This term has the smallest positive value so
0	1	0	0	$-\frac{1}{4}$	$\frac{1}{4}$	6	$\leftarrow R_4-\frac{1}{5}R_3$	now manipulate the matrix to express P in terms of r and s)
1	0	0	0	$\frac{5}{4}$	$-\frac{1}{4}$	10		
0	0	0	$\frac{4}{3}$	$-\frac{19}{3}$	1	8	$\leftarrow R_2/\left(\frac{3}{2}\right)$	
0	0	1	0	$\frac{5}{4}$	$-\frac{1}{4}$	2		
0	1	0	0	$-\frac{1}{4}$	$\frac{1}{4}$	6		
1	0	0	$\frac{1}{3}$	$\frac{2}{3}$	0	12	$\leftarrow R_1+\frac{1}{4}R_2$	From the first row P could now be expressed in terms of r and s, with a negative coefficient
0	0	0	$\frac{4}{3}$	$-\frac{19}{3}$	1	8		of each, so now stop; i.e. since the top row has all positive coefficients you can see that the
0	0	1	$\frac{1}{3}$	$-\frac{1}{3}$	0	4	$\leftarrow R_3+\frac{1}{4}R_2$	maximum value of P is 12 and that it is reached when $r=s=0$ (which happens when
0	1	0	$-\frac{1}{3}$	$\frac{4}{3}$	0	4	$\leftarrow R_4-\frac{1}{4}R_2$	$x=4$ and $y=4$).

The advantage of this method is that it can be readily extended to problems with more than two variables, as shown below

Example

Maximise $P = 4x + 5y + 3z$

subject to $8x + 5y + 2z \leq 3$

$$3x + 6y + 9z \leq 2$$

$$x, y, z \geq 0$$

Solution

As usual slack variables r and s are introduced;

$$8x + 5y + 2z + r = 3$$
$$3x + 6y + 9z + s = 2$$

Now x, y, z, r, $s \geq 0$ and the simplex tableau is shown below

P	x	y	z	r	s		Comments
1	-4	-5	-3	0	0	0	Increase x initially and compare $3/8^*$, $2/3$. This smaller value of x occurs where $y=z=r=o$ and so now manipulate the matrix to express P in terms of y, z and r.
0	8	5	2	1	0	3	
0	3	6	9	0	1	2	
1	-4	-5	-3	0	0	0	
0	1	$\frac{5}{8}$	$\frac{2}{8}$	$\frac{1}{8}$	0	$\frac{3}{8}$	$\leftarrow R_2/8$
0	3	6	9	0	1	2	
1	0	$-\frac{5}{2}$	-2	$\frac{1}{2}$	0	$\frac{3}{2}$	$\leftarrow R_1 + 4R_2$
0	1	$\frac{5}{8}$	$\frac{2}{8}$	$\frac{1}{8}$	0	$\frac{3}{8}$	Increase y and compare $(3/8)/(5/8)$, $(7/8)/(33/8)^*$. This smaller value occurs when $z=r=s=o$ and so express P in terms of z, r and s.
0	0	$\frac{33}{8}$	$\frac{33}{4}$	$-\frac{3}{8}$	1	$\frac{7}{8}$	$\leftarrow R_3 - 3R_2$
1	0	$-\frac{5}{2}$	-2	$\frac{1}{2}$	0	$\frac{3}{2}$	
0	1	$\frac{5}{8}$	$\frac{2}{8}$	$\frac{1}{8}$	0	$\frac{3}{8}$	
0	0	1	2	$-\frac{1}{11}$	$\frac{8}{33}$	$\frac{7}{33}$	$\leftarrow R_3 / \left(\frac{33}{8}\right)$
1	0	0	3	$\frac{3}{11}$	$\frac{20}{33}$	$\frac{67}{33}$	$\leftarrow R_1 + \frac{5}{2}R_3$ — The first row now has positive coefficients, showing that there is a maximum of $67/33$ when $z=r=s=o$ (which happens when $x=8/33$, $y=7/33$ and $z=0$).
0	1	0	-1	$\frac{2}{11}$	$-\frac{5}{33}$	$\frac{8}{33}$	$\leftarrow R_2 - \frac{5}{8}R_3$
0	0	1	2	$-\frac{1}{11}$	$\frac{8}{33}$	$\frac{7}{33}$	

Exercise 5D

Use the simplex algorithm to solve the following problems.

1. Maximise $P = 4x + 6y$

   subject to $x + y \leq 8$

   $\qquad\qquad 7x + 4y \leq 14$

   $\qquad\qquad\quad x \geq 0$

   $\qquad\qquad\quad y \geq 0$

2. Maximise $P = 10x + 12y + 8z$

   subject to $2x + 2y \leq 5$

   $\qquad\qquad 5x + 3y + 4z \leq 15$

   $\qquad\qquad\quad x \geq 0$

   $\qquad\qquad\quad y \geq 0$

   $\qquad\qquad\quad z \geq 0$

3. Maximise $P = 3x + 8y - 5z$

   subject to $2x - 3y + z \leq 3$

   $\qquad\qquad 2x + 5y + 6z \leq 5$

   $\qquad\qquad\quad x \geq 0$

   $\qquad\qquad\quad y \geq 0$

   $\qquad\qquad\quad z \geq 0$

4. Maximise $3x + 6y + 2z$

   subject to $3x + 4y + 2z \leq 2$

   $\qquad\qquad x + 3y + 2z \leq 1$

   $\qquad\qquad\quad x \geq 0$

   $\qquad\qquad\quad y \geq 0$

   $\qquad\qquad\quad z \geq 0$

5.5   Miscellaneous Exercises

1. Find the solution to Question 3 of Exercise 5A.

2. A firm manufactures two types of box, each requiring the same amount of material.

   They both go through a folding machine and a stapling machine.

   Type A boxes require 4 seconds on the folding machine and 3 seconds on the stapling machine.

   Type B boxes require 2 seconds on the folding machine and 7 seconds on the stapling machine.

   Each machine is available for 1 hour.

   There is a profit of 40p on Type A boxes and 30p on Type B boxes.

   How many of each type should be made to maximise the profit?

3. A small firm which produces radios employs both skilled workers and apprentices. Its workforce must not exceed 30 people and it must make at least 360 radios per week to satisfy demand. On average a skilled worker can assemble 24 radios and an apprentice 10 radios per week.

   Union regulations state that the number of apprentices must be less than the number of skilled workers but more than half of the number of skilled workers.

   What is the greatest number of skilled workers than can be employed?

   Skilled workers are paid £300 a week, and unskilled workers £100 a week.

   How many of each should be employed to keep the wage bill as low as possible?

4. In laying out a car park it is decided, in the hope of making the best use of the available parking space (7200 sq.ft.), to have some spaces for small cars, the rest for large cars. For each small space 90 sq.ft. is allowed, for each large space 120 sq.ft. Every car must occupy a space of the appropriate size. It is reliably estimated that, of the cars wishing to park at any given time, the ratio of small to large will be neither less that 2:3 nor greater than 2:1.

   Find the number of spaces of each type in order to maximise the number of cars that can be parked.

*5. A contractor hiring earth moving equipment has the choice of two machines.

   Type A costs £25 per day to hire, needs one man to operate it and moves 30 tonnes of earth per day.

   Type B costs £10 per day to hire, needs four men to operate it and moves 70 tonnes of earth per day.

   The contractor can spend up to £500 per day, has a labour force of 64 men available and can use a maximum of 25 machines on the site.

   Find the maximum weight of earth that the contractor can move in one day.

6. A landscape designer has £200 to spend on planting trees and shrubs to landscape an area of 1000 m². For a tree he plans to allow 25 m² and for a shrub 10 m². Planting a tree will cost £2 and a shrub £5.

   If he plants 30 shrubs what is the maximum number of trees he can plant?

   If he plants 3 shrubs for every tree, what is the maximum number of trees he can plant?

7. A small mine works two coal seams and produces three grades of coal. It costs £10 an hour to work the upper seam, obtaining in that time 1 tonne of anthracite, 5 tonnes of best quality coal and 2 tonnes of ordinary coal. The lower seam is more expensive to work, at a cost of £15 per hour, but it yields in that time 4 tonnes of anthracite, 6 tonnes of best coal and 1 tonne of ordinary coal. Faced with just one order, for 8 tonnes of anthracite, 30 tonnes of best coal and 8 tonnes of ordinary coal each day, how many hours a day should each seam be worked so as to fill this order as cheaply as possible?

6 PLANAR GRAPHS

Objectives

After studying this chapter you should

* be able to use tests to decide whether a graph is planar;

* be able to use an algorithm to produce a plane drawing of a planar graph;

* know whether some special graphs are planar;

* be able to apply the above techniques and knowledge to problems in context.

6.0 Introduction

This topic is introduced through an activity.

Activity 1

A famous problem is that of connecting each of three houses, as shown opposite, to all three services (electricity, gas and water) with no pipe/cable crossing another.

Try this problem. Four of the nine lines needed have already been put into the picture.

Investigate the problem for different numbers of houses and services.

What happens if the scene is on the surface of a sphere (which in reality it is) or on a torus (a ring doughnut!) or on a Möbius strip?

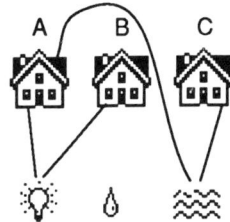

You will have found graphs which can be completed without their edges crossing and some graphs which cannot. If a graph can be drawn in the plane (on a sheet of paper) without any of its edges crossing, it is said to be **planar**.

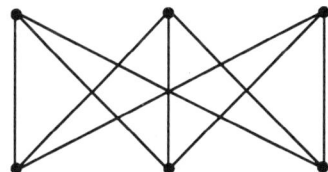

The graph shown on the right is planar, although you might not think so from the first diagram of it. The next two diagrams are of the same graph and confirm that it is planar. This diagram is called **a plane drawing** of the graph.

As you should have found in the activity above, the graph shown on the previous page, which represents the services problem, cannot in fact be drawn without crossing edges and is therefore described as **non-planar**. This has repercussions for the electronics industry, because it means that a simple circuit with six junctions and wires connecting each of three junctions to each of the other three junctions cannot be made without cross-overs. In an integrated circuit within a 'chip', this would mean two 'layers' of wires.

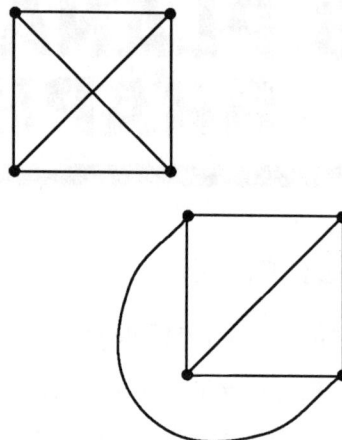

6.1   Plane drawings

Activity 2

Try re-drawing the two graphs shown on the right so that no edges cross.

The second of the two graphs is called K_5, the complete graph with five vertices: each vertex is joined to every other one by an edge. Of course, $K_6, K_7, ...$, are similarly defined. Although K_5 looks simpler than the one shown above it, it is in fact non-planar, whereas the one above it is planar.

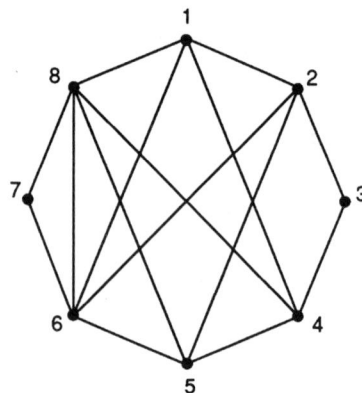

Later on it will be proved that both K_5 and another graph called $K_{3,3}$ (which is the one associated with the gas, water and electricity problem in Activity 1) are non-planar, and you will see the significance of this when you look at Kuratowski's Theorem later in this chapter.

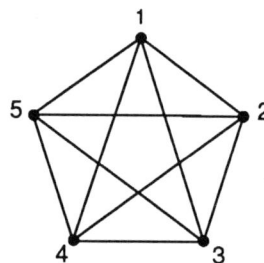

Exercise 6A

1. Sketch the graphs of K_4 and K_6. Are they planar? For which values of n do you think K_n is planar?

2. For which values of positive integers is K_n Eulerian? For which is it semi-Eulerian?

3. There are five so-called Platonic solids with a very regular structure Graphs based on the first three are shown below.

   Make plane drawings of each, if possible.

   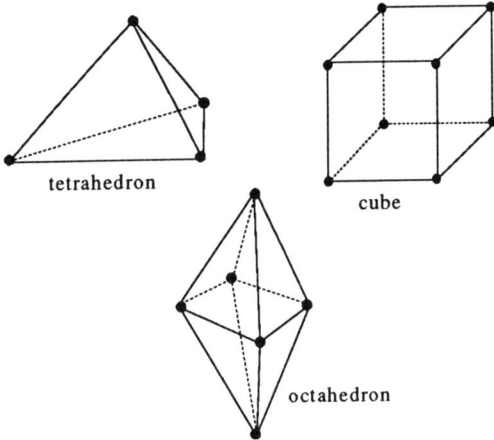

   tetrahedron

   cube

   octahedron

4. The graph K_n can be used to represent the games played during a 'round robin' tournament in which each player plays every other player.

   How many games take place when there are

   (a)   5 players

   (b) n players?

5. Draw a graph with six vertices, labelled 1 to 6, in which two vertices are joined by an edge if, and only if, they are **co-prime** (i.e. if they have no common factor greater than 1). Is the graph planar?

6.2   Bipartite graphs

The graph associated with the activity in Section 6.1 is called a **bipartite** graph. Such graphs consist of two sets of vertices, with edges only joining vertices between sets and not within a set. The diagrams opposite are of bipartite graphs. In the second one the two sets of vertices contain three vertices and two vertices and every vertex in the first set is joined to every vertex in the second: this graph is called $K_{3,2}$ (and, of course, $K_{r,s}$ could be defined similarly for any positive integers r and s).

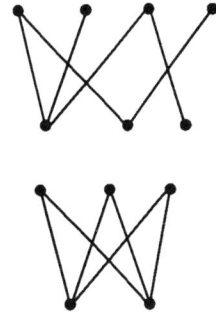

Exercise 6B

1. Sketch $K_{3,4}$ and $K_{4,2}$.
2. How many edges are there in general in the graph $K_{r,s}$?

3. Two opposing teams of chess players meet for some games. Show how a bipartite graph can be used to represent the games actually played. If the graph turned out to be $K_{r,s}$ what would it mean?

4. For which r and s is $K_{r,s}$ Eulerian? For which r and s is it semi-Eulerian?

6.3   A planarity algorithm

Naturally, for very complicated graphs it would be convenient to have a technique available which will tell you both whether a graph is planar and how to make a plane drawing of it.

Activity 3

Using the first graph shown opposite as an example, try to develop an algorithm in order to construct a planar graph.

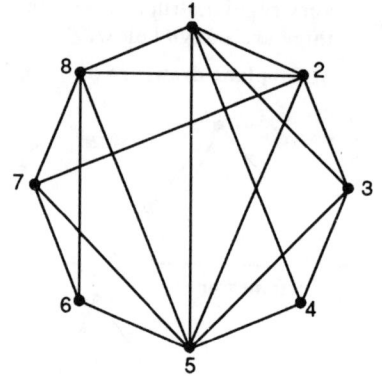

The algorithm described below can be applied only to graphs which have a Hamiltonian cycle; that is, where there is a cycle which includes every vertex of the graph.

The method will be illustrated by applying it to the graph shown in Activity 3, above.

The first stage is to redraw the graph so that the Hamiltonian cycle forms a regular polygon and all edges are drawn as straight lines inside the polygon. The graph used here is already in this form, but for other graphs this stage might involve 'moving' vertices as well as edges.

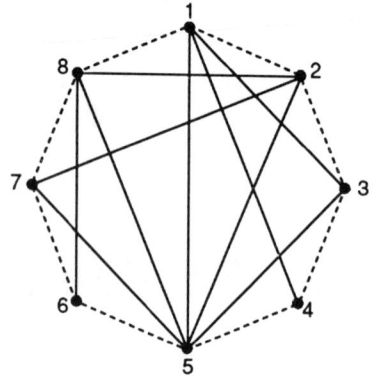

The edges of the regular polygon now become part of the solution (shown dotted in the second graph).

The next stage is to choose any edge, say 1 - 3, and decide whether this is to go inside or outside: let's choose inside, as illustrated in the third graph opposite.

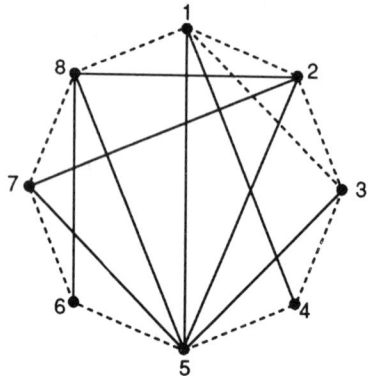

Since 1 - 3 crosses 2 - 8, 2 - 7 and 2 - 5, all these edges must go outside as shown.

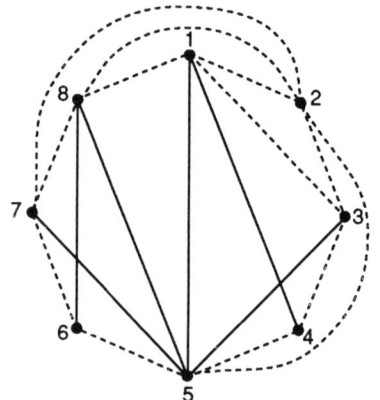

Originally edge 2 - 7 crossed 1 - 4, 1 - 5, 8 - 5 and 8 - 6 , so all these edges must now remain inside (or they would cross 2 - 7 outside).

Finally, because 1 - 4 stays inside, 3 - 5 must go outside, and since 8 - 6 stays inside, 7 - 5 must also go outside, as shown.

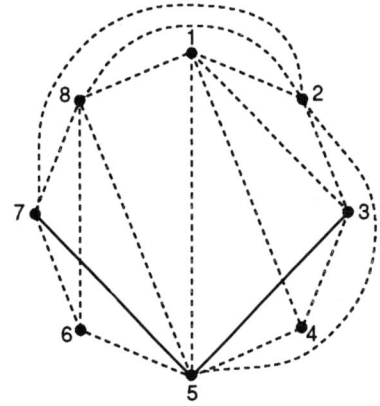

This is now a planar graph, as shown opposite, where the dotted lines have been redrawn as solid lines.

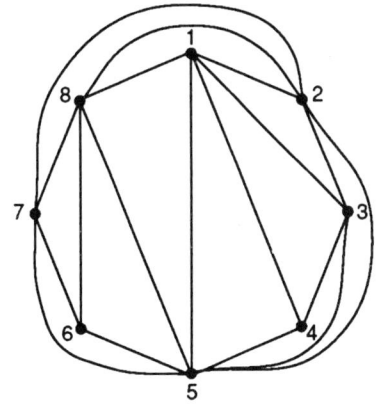

The method illustrated above can also be used to show whether or not a graph is planar. For example, consider K_5, and, as before, the regular polygon is first included as part of the solution.

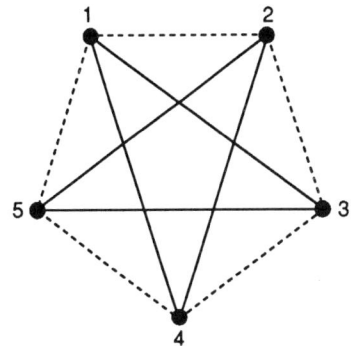

Choose an edge, say 1 - 3, which stays inside. Since this crosses 2 - 5 and 2 - 4, both of these will have to go outside as shown.

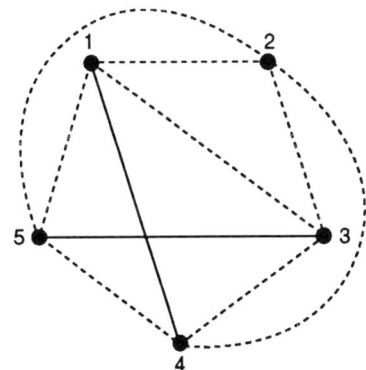

Now 2 - 5 crosses 1 - 4, so 1 - 4 must stay inside, as shown.

Finally, consider edge 3 - 5. Since it crosses 1 - 4, it must go outside; but it also crosses 2 - 4 which is already outside; so 3 - 5 must also go outside! This is a contradiction and it is concluded that the graph is non-planar.

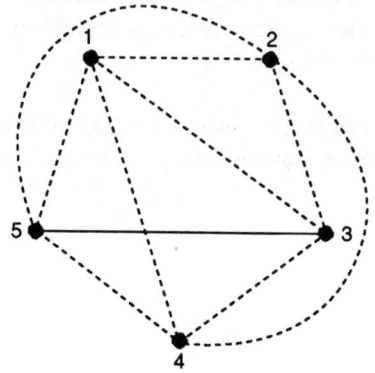

Example

Use the planarity algorithm to find a plane drawing of the graph opposite.

Solution

The graph has a Hamiltonian cycle

$$1 - 2 - 3 - 4 - 5 - 6 - 7 - 8 - 9 - 10 - 11 - 12 - 13 - 14 - 1$$

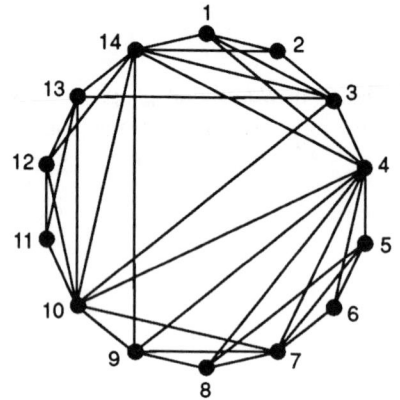

which is part of the solution, as indicated by the dotted lines in the second graph.

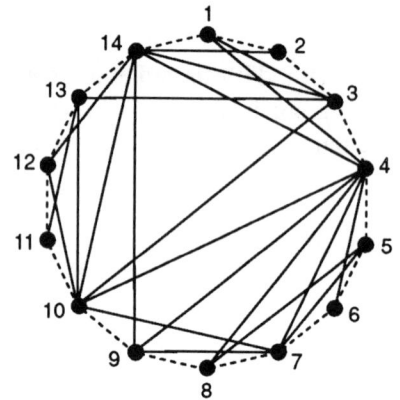

Choose any edge, say 13 - 3, and keep it in place (shown dotted). All edges that cross this line must now be put outside.

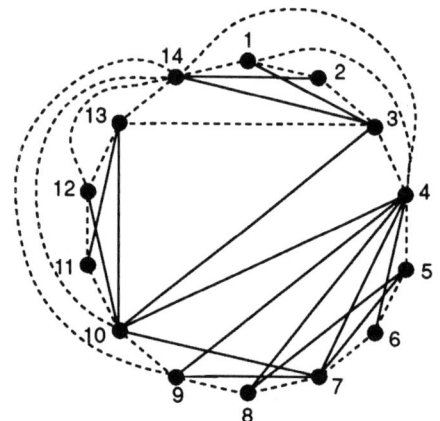

Since 1 - 4 crosses 14 - 3 and 14 - 2, they must go inside, and similarly 1 - 3 must go outside.

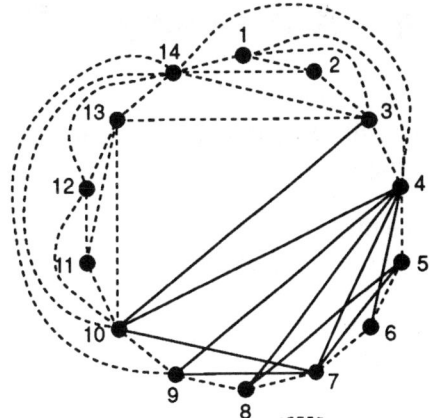

Also, 14 - 4 crosses 10 - 3 so 10 - 3 must stay inside; and since 14 - 9 is now outside, 10 - 4 and 10 - 7 must stay inside. Also 14 - 12 outside implies 13 - 10 and 13 - 11 inside, which then means that 12 - 10 must be drawn outside (as shown opposite).

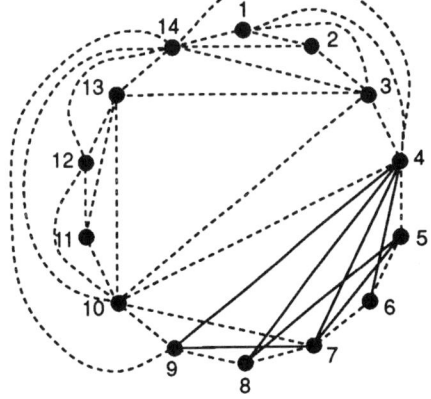

Continuing in this way, 10 - 7 inside means that 9 - 4, 8 - 4, 8 - 5 all go outside; which in turn means that 9 - 7 , 7 - 4 and 6 - 4 go inside. 7 - 5 must go outside.

You now have a plane drawing of the graph as shown opposite; the lines are not dotted since this is the final solution.

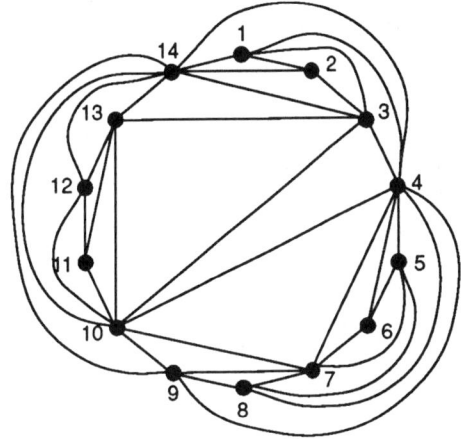

Is this solution unique?

Exercise 6C

1. Make plane drawings of the following two graphs

2. Show that the graphs of K_6 and $K_{3,3}$ are not planar by using the algorithm. (Note that $K_{3,3}$ must first be redrawn to form a regular polygon)

3. Show that this Petersen graph is non-planar.

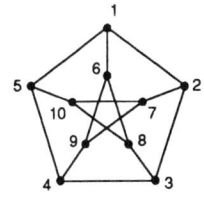

4. By first redrawing with a regular polygon, use the planarity algorithm to produce a plane drawing of the graph shown opposite.

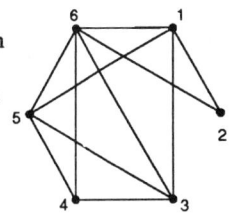

6.4   Kuratowski's Theorem

The non-planar graphs K_5 and $K_{3,3}$ seem to occur quite often. In fact, all non-planar graphs are related to one or other of these two graphs.

To see this you first need to recall the idea of a **subgraph**, first introduced in Chapter 1 and define a **subdivision** of a graph.

A subgraph is simply a part of a graph, which itself is a graph. G_1 is a subgraph of G as shown opposite.

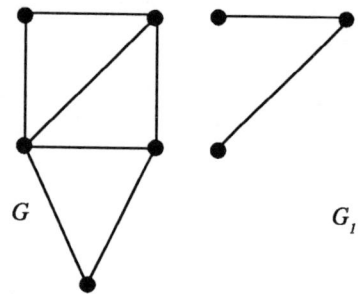

A subdivision of a graph is the original graph with added vertices of degree 2 along the original edges. As shown opposite, G_2 is a subdivision of G.

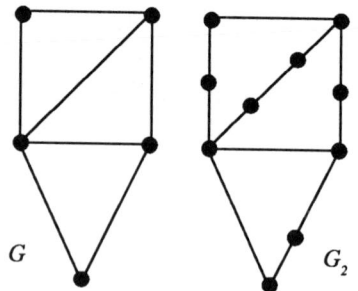

When a planar graph is subdivided it remains planar; similarly if it is non-planar, it remains non-planar.

Kuratowski's Theorem states that a graph is planar if, and only if, it does not contain K_5 and $K_{3,3}$, or a subdivision of K_5 or $K_{3,3}$ as a subgraph.

This famous result was first proved by the the Polish mathematician *Kuratowski* in 1930. The proof is beyond the scope of this text, but it is a very important result.

The theorem will often be used to show a graph is non-planar by finding a subgraph of it which is either K_5 or $K_{3,3}$ or a subdivision of one of these graphs.

Example

Graph G has been redrawn, omitting edges 3 - 6 and 4 - 6. Thus G' is a subgraph of G.

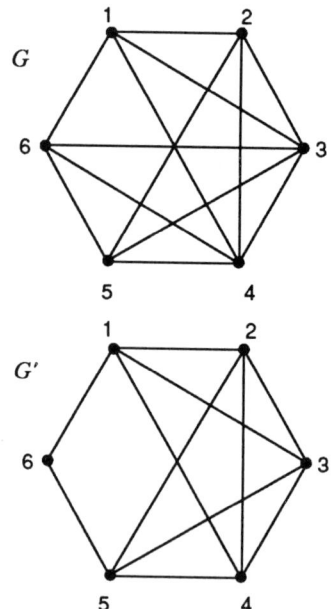

Also G' without vertex 6 is isomorphic to K_5. The addition of vertex 6 makes G' a subdivision of K_5. So G', a subdivision of K_5, is a subgraph of G, and therefore G is non-planar.

Exercise 6D

1. Which of these graphs are subdivisions of $K_{3,3}$ and why?

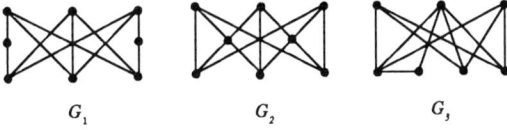

G_1 G_2 G_3

2. Use Kuratowski's Theorem to show that the following graphs are non-planar.

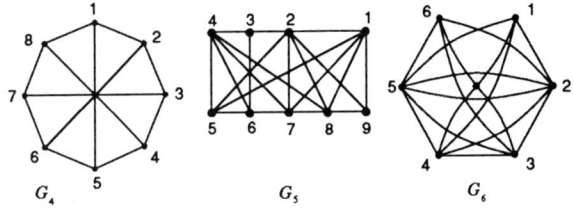

G_4 G_5 G_6

6.5   Miscellaneous Exercises

1. Give examples of
   (a) a planar graph in which each vertex has degree 4, and
   (b) a planar graph with six vertices and a shortest cycle of length 4.

2. For which values of r, s is the complete bipartite graph $K_{r,s}$ non-planar?

3. The crossing number of a graph is the least number of points at which edges cross. What are the crossing numbers of
   (a) $K_{3,3}$
   (b) K_6
   (c) $K_{1,2}$?

4. Use Kuratowski's Theorem in order to prove that the graph G is non-planar.

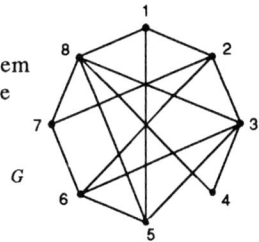

G

5. Show that this graph is non-planar.

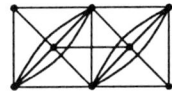

7 NETWORK FLOWS

Objectives

After studying this chapter you should

- be able to draw network diagrams corresponding to flow problems;

- be able to interpret networks;

- be able to find optimum flow rates in a network, subject to constraints;

- be able to use the labelling algorithm to find the maximum flow rate in a network;

- be able to interpret the analysis of a network for real life problems.

7.0   Introduction

There are many situations in life which involve flow rates; some are self-evident, such as traffic flow or the flow of oil in a pipeline; others have the same basic structure but are less obviously flow problems - e.g. movement of money between financial institutions and activity networks for building projects. In most of the problems you will meet, the objective is to maximise a flow rate, subject to certain constraints. In order to get a feel for these types of problem, try the following activity.

Activity 1

This diagram represents a road network. All vehicles enter at S and leave at T. The numbers represent the maximum flow rate in vehicles per hour in the direction from S to T. What is the maximum number of vehicles which can enter and leave the network every hour?

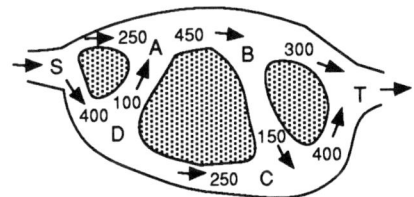

Which single section of road could be improved to increase the traffic flow in the network?

7.1 Di-graphs

The network in the previous activity can be more easily analysed when drawn as a graph, as shown opposite.

The arrows show the flow direction; consequently this is called a **directed graph** or **di-graph**. In this case the edges of the graph also have **capacities** : the maximum flow rate of vehicles per hour. The vertices S and T are called the **source** and **sink**, respectively.

You should have found that the maximum rate of flow for the network is 600. This is achieved by using each edge with flows as shown.

Notice that some of the edges are up to maximum capacity, namely SA, BT, DA and DC. These edges are said to be **saturated**. Also, at any vertex, other than S or T, in an obvious sense, the **inflow** equals the **outflow**.

Di-graphs for some situations show no capacities on the edges. For example, suppose you have a **tournament** in which four players each play one another. If a player A beats a player B then an arrow points from A to B. In the diagram opposite you can see that A beats D, but loses to B and C.

In what follows, the term **network** will be used to denote a directed graph with capacities.

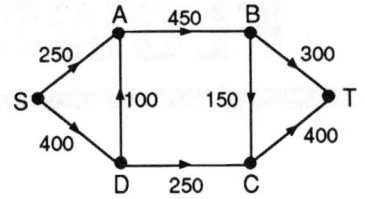

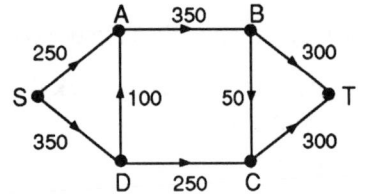

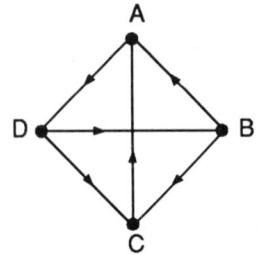

Exercise 7A

1. The diagrams below show maximum flow capacities in network N_1, and actual intended flows in N_2.

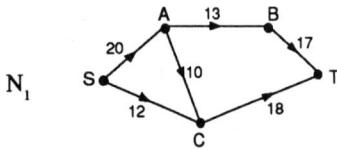

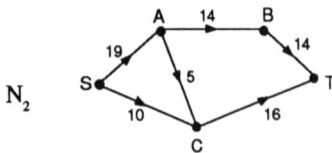

   What errors have been made in constructing N_2? Draw a new network which has a maximum flow from S to T.

2. Draw a network representing the results in the tournament described by this table.

	A	B	C	D	E
A	•	X	O	X	O
B	O	•	(X)	X	X
C	X	O	•	O	X
D	O	O	X	•	X
E	X	O	O	O	•

X denotes a win. O denotes a loss.

For example, the (X) shows that B beats C.

7.2 Max flow - min cut

The main aim is to find the **value** of the maximum **flow** between the source and sink. You will find the concept of the **capacity** of a **cut** very useful. The network opposite illustrates a straightforward flow problem with maximum allowable flows shown on the edges.

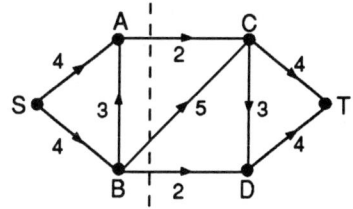

The dotted line shown in the first diagram illustrates one possible **cut**, which separates S from T. Its **capacity** is defined as the sum of the maximum allowable flows across the cut; i.e. $2+5+2=9$. There are many possible cuts across the network. Two more are shown in the second diagram. For L_1, the capacity is

$$2+0+4=6.$$

The reason for the zero is as follows: the flows in AC and SB cross the cut from left to right, whereas the flow in BA crosses from right to left. To achieve maximum flow across the cut the capacity of BA is not used.

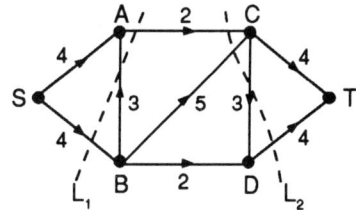

Similarly for L_2, the capacity is given by

$$2+5+0+4=11.$$

Activity 2

For the network shown above, find all possible cuts which separate S from T, and evaluate the capacity of each cut. What is the minimum capacity of any cut?

What do you notice about the capacities?

Activity 3

Find the maximum flow for the network shown above. What do you notice about its value?

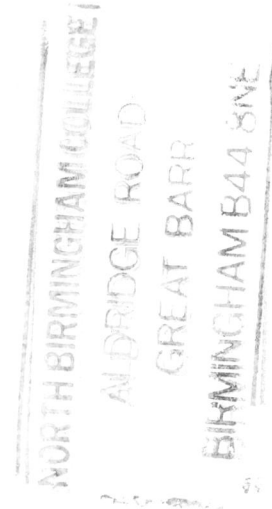

The activities above give us a clue to the max flow-min cut theorem. You should have noticed that the maximum flow found equals the cut of minimum capacity. In general,

value of any flow $\leq$ capacity of any cut

and equality occurs for maximum flow and minimum cut; this can be stated as

maximum flow = minimum cut.

Example

For the network opposite, find the value of the maximum flow and
a cut which has capacity of the same value.

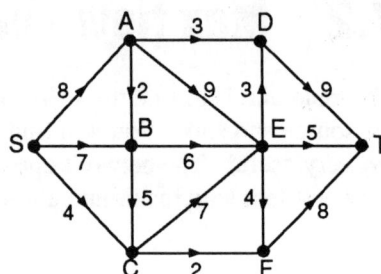

Solution

By inspection, the maximum flow has value 17; this is illustrated
by the circled numbers on the network opposite. Also shown is a
cut of the same capacity.

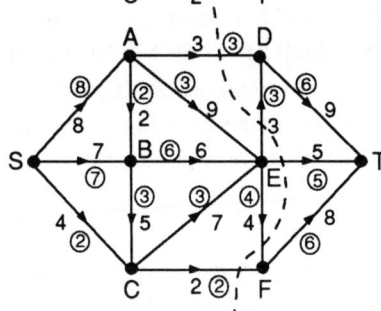

*If you can find a cut and flow of the same value, can you be sure
that you have found the maximum flow?*

Exercise 7B

1. The network below shows maximum capacities
   of each edge. Draw up a table showing the
   values of all the cuts from A, B to C, D, E.
   Which is the minimum cut? Draw the network
   with flows which give this maximum total flow.

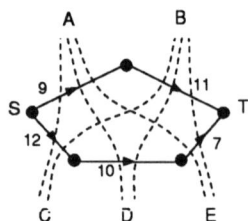

2. Find a minimum cut for each of these networks.
   The numbers along the edges represent
   maximum capacities.

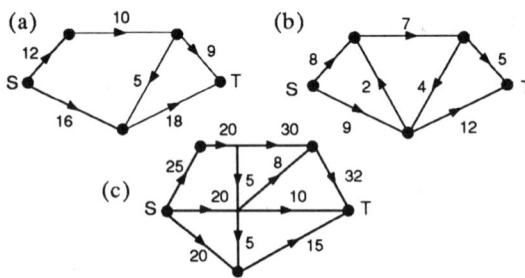

3. For each of the networks in question 2 try to find
   values for the flows in each edge which give the
   maximum overall flow.

7.3   Finding the flow

You may have noticed that the minimum cut is coincident with
edges which have a flow equal to their maximum capacity.

The diagram opposite shows a network with its allowable
maximum flow along each edge. The minimum cut is marked L.
It has a capacity of 15. This line cuts the edges with capacities 7
and 8. The actual maximum flow of value 15 is shown in the
diagram, and it should be noted that the minimum cut only passes
through edges that are saturated (or have zero flow in the opposing
direction).

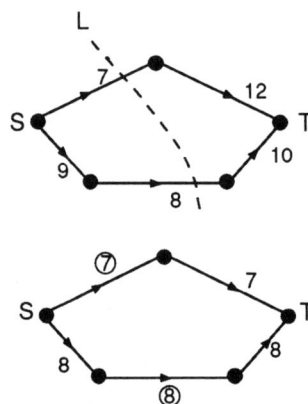

This information should help you to confirm maximum flows. Note that in some cases there is more than one possible pattern for the flows in the edges which give the overall maximum flow.

Activity 4

By trial and error, find the maximum possible flow for the network opposite.

Find a cut which has a capacity equal to the maximum flow (you might find it helpful to mark each edge which is satisfied by the maximum flow - the minimum cut will only cut saturated edges or edges with zero flows in the opposing direction.)

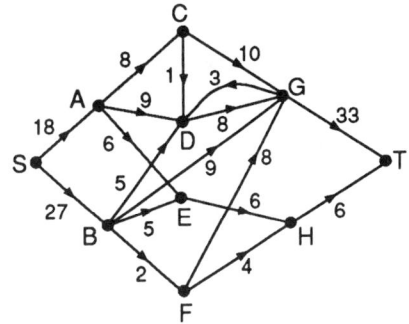

Exercise 7C

1. Find the maximum flow for each of these networks, and show the minimum cut in each case.

(a)

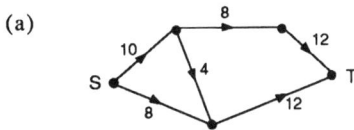

(b)

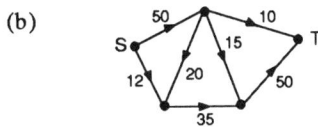

(c)

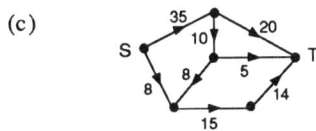

2. A network has edges with maximum capacities as shown in this table.

	S	A	B	C	D	E	T
S	•	40	40	•	•	•	
A	•	•	•	•	15	20	
B	•	•	•	45	•	•	
C	•	•	•	•	•	•	50
D	•	•	10	15	•	•	15
E	•	•	•	•	•	•	25

The letters refer to vertices of the network, where S and T are the source and sink respectively.

Draw a diagram of the network.

Find a maximum flow for the network, labelling each edge with its actual flow.

7.4 Labelling flows

So far you have no method of actually finding the maximum flow in a network, other than by intuition.

The following method describes an algorithm in which the edges are labelled with artificial flows in order to optimise the flow in each arc.

An example follows which shows the use of the **labelling algorithm**.

Example

The network opposite has a maximum flow equal to 21, shown by the cut XY. When performing the following algorithm you can stop, either when this maximum flow has been reached or when all paths from S to T become 'saturated'.

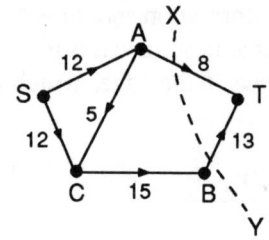

---

1. Note that there are four possible paths from S to T, namely SAT, SCBT, SCAT, SACBT.

2. Begin with any of these, say SAT, as in the diagram opposite. The maximum flow is restricted by AT, so label each edge with its **excess capacity**, given that AT carries its marked capacity, as shown.

3. Both flows could be reduced by up to 8 (the capacity of edge AT). Show it as a potential backflow in each edge.

4. Now add this section back on to the original network as shown and choose another route, say SACT.

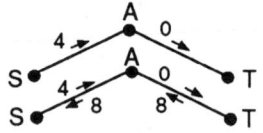

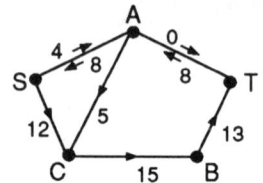

   Of the possible flows, $S \to A$, $A \to C$ and $C \to T$, note that the lowest is 4 and this represents the maximum flow through this path, as shown.

5. As before, each edge in the path SACT is labelled with its excess capacity (above 4), and the reverse flows, noting that the sum of the forward and reverse flows always equals the original flow, as shown opposite. Note particularly that the excess flow in SA has now dropped to zero.

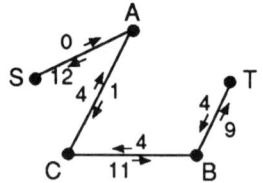

   The resulting network is shown opposite.

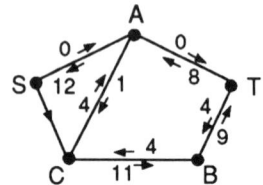

6. Continue by choosing a third path, say SCBT, and inserting artificial forward and backward excess flows.

   The network is shown opposite.

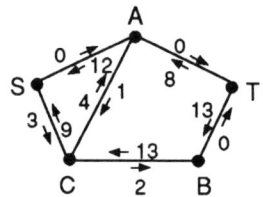

   There is one more route, but it is unnecessary to proceed with the process because the flows to T from A and B are saturated, shown by zero excess flow rates. This means that the flow can increase no further.

7. The excess flows can be subtracted from the original flows to create the actual flows or you can simply note that the back flows give the required result - but with the arrows reversed. The final result is shown opposite.

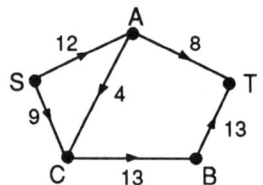

The method looks quite complicated, but after a little practice you should become quite adept at it.

Exercise 7D

1. Use the labelling algorithm in order to find the maximum flow in each of these networks, given the maximum capacity of each edge.

   (a)

   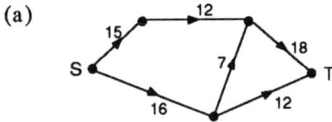

   (b)

   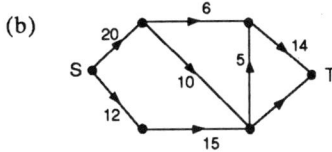

   (c)

   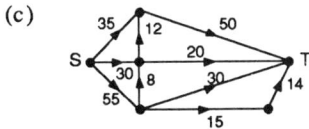

2. There are a number of road routes from town A to town B as shown in the diagram below. The numbers show the maximum flow rate of vehicles in hundreds per hour. Find the maximum flow rate of vehicles from A to B. Suggest a single road section which could be widened to improve its flow rate. How does this affect traffic flow on other sections, if the network operates to its new capacity?

   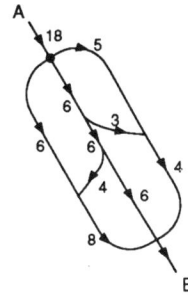

7.5 Super sources and sinks

Many networks have multiple sources and/or sinks. A road network with two sources and three sinks is shown opposite.

The problem of finding the maximum flow can be quite easily dealt with by creating a single **super source** S and a single **super sink** T.

The resulting network is as shown and the usual methods can now be applied.

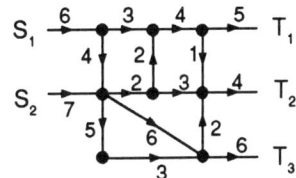

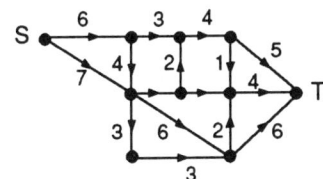

Activity 5

Add a super source and a super sink to this network, in which maximum capacities are shown, and then use the labelling algorithm to find a maximum flow through it.

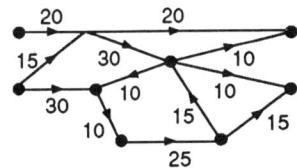

Activity 6

Investigate the traffic flow in a small section of the road network near to you, for which you could estimate maximum flows in each road.

7.6 Minimum capacities

Sometimes edges in networks also have a minimum capacity
which has to be met. In the diagram opposite, for example, edge
AB has a maximum capacity of 6 and a minimum of 4. The flow
in this edge must be between 4 and 6 inclusive.

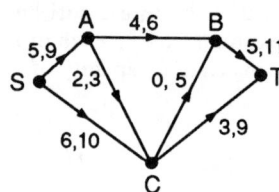

Activity 7

Find the maximum flow in the network shown above. Investigate
how the max flow - min cut theorem can be adapted for this
situation.

Example

Find the maximum flow for the network shown opposite.

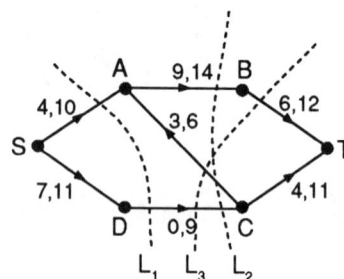

Solution

In order to find a minimum cut, the max flow - min cut theorem is
adapted so that you **add** upper capacities of edges along the cut
directed from S to T, but **subtract** lower capacities of edges
directed from T to S.

For this network, cut L_1 has a value $10 + 9 = 19$, but cut L_2 has a
value of $14 + 9 - 3 = 20$, since edge CA crosses L_2 from T to S. In
fact, L_3 is the minimum cut - with a value of $12 - 3 + 9 = 18$, so you
are looking for a maximum flow of 18.

If you are familiar with the labelling algorithm, here is a slightly
quicker version .

1. Begin with **any** flow. The one shown opposite will do.
    Note that none of the upper or lower capacities of the edges
    has been violated. It is not the best because the flow is
    only 15.

2. For each edge, insert the potential excess flow and the
    corresponding back flow. For example, BT carries a flow
    of 10 at present. It could be 2 more and it could be 4 less,
    since the minimum and maximum flows in the edge are
    6 and 12.

    This has been done for all the edges in the network
    resulting in the diagram opposite.

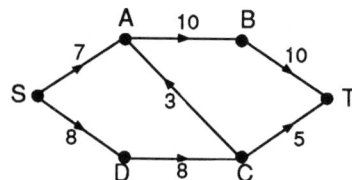

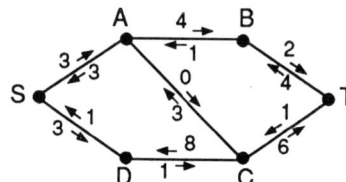

3. Now look for a path in which the flow (from S to T) can be improved. Consider SABT. The lowest excess capacity in these three edges is 2 (in BT) so the flow in each edge can be improved by this amount.

    SABT is called a **flow augmenting path** because its overall excess flow can be reduced. The reverse flow has to be increased by 2 to compensate. The path now looks like this.

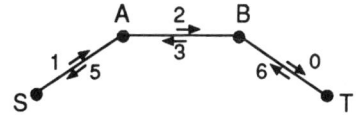

4. This can now be added to the network as shown opposite.

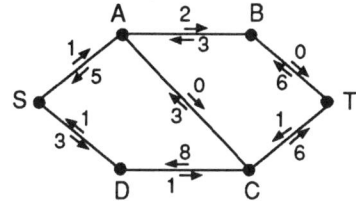

5. Now look for another path which can be augmented (improved).

    SACT cannot, since AC has a flow of zero from S to T.

    SDCABT cannot, since BT has a zero flow. The only possibility is SDCT. This can be improved by an increase of one. The result is shown here.

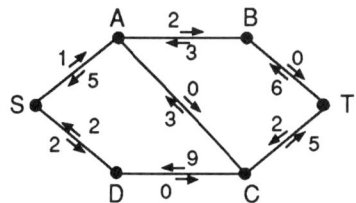

6. Now subtract the excess (S → T directed) flows from their maximum values. So, for example, AB becomes $14 - 2 = 12$. The final network - as shown opposite - has a flow of 18 as required.

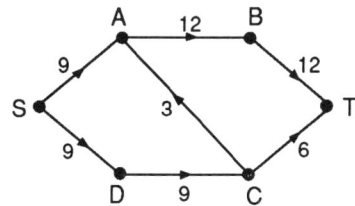

Exercise 7E

1. For the network of Activity 7 find any flow in the network and then use the labelling algorithm to find the maximum flow.

2. Which, if any, of the following networks showing upper and lower capacities of edges has a possible flow? If there is no possible flow, explain why.

    (a)
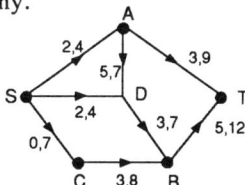

    (b)
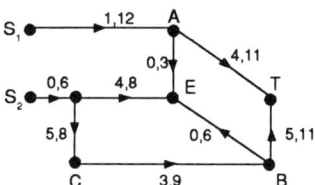

3. Find a maximum flow from S to T in this network showing upper and lower capacities.

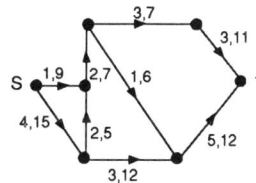

7.7 Miscellaneous Exercises

1. Find a minimum cut for each of the following networks.

   In N_1 and N_3 edges can carry the maximum capacities shown. In N_2 the minimum and maximum capacities of the edge are shown.

   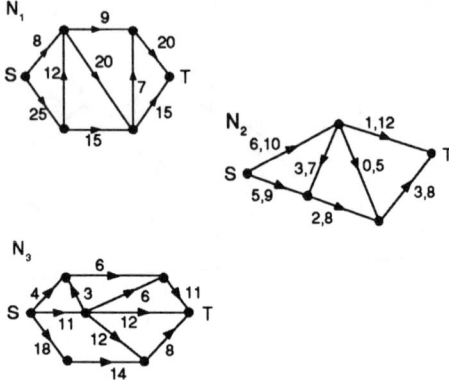

2. Use the labelling algorithm to find a maximum flow in this network, which shows maximum capacities.

   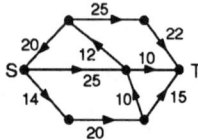

3. Which of these values of x and y gives a possible flow for the networks shown below with upper and lower capacities?

   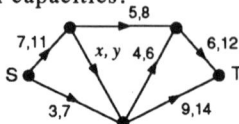

   (a) 7, 10   (b) 3, 12

   (c) 4, 5   (d) 1, 8

   For these cases find a maximum flow for the networks.

4. By creating a super source and a super sink find a possible flow for this network which shows maximum capacities.

   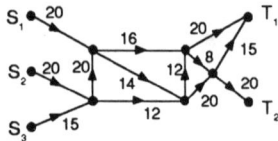

5. The table shows the daily maximum capacity of coaches between various cities (in hundreds of people).

   Draw a network to show the capacities of the routes from London through to Newcastle.

   A festival is taking place in Newcastle. Find the maximum number of people who can travel by coach from London for the festival. Investigate what happens when there is a strike at one of the coach stations, say Liverpool.

From \ To	Lon	Bir	Man	Lds	Lpl	New
London	•	40	•	20	•	•
Birmingham	•	•	10	15	12	•
Manchester	•	•	•	12	•	15
Leeds	•	•	•	•	•	30
Liverpool	•	•	7	•	•	8

6. Find a maximum traffic flow on this grid-type road system from X to Y, in which maximum flow rates are given in hundreds of vehicles per hour.

   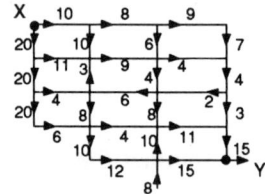

7. Find the maximum flow through this network showing maximum capacities.

   Use a die to simulate flow through the network for each section every hour. What happens?

   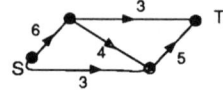

8. The following underground map shows a 'circular route' with 8 stations. Trains travel only in the direction shown. The capacities indicate the maximum number of trains per hour which can pass along each section. **At least** one train per hour must travel along each section of track. A train can carry 500 passengers. Find the maximum number of passengers which can flow from A to B.

   Note that A and B are not sources or sinks. The number of trains in the system must always remain constant.

   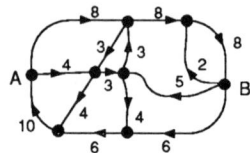

8 LOGIC

Objectives

After studying this chapter you should

- understand the nature of propositional logic;
- understand the connectives NOT, OR, AND;
- understand implication and equivalence;
- be able to use truth tables;
- be able to identify tautology and contradiction;
- be able to test the validity of an argument.

8.0   Introduction

It may seem unusual for philosophical ideas of logic based on intuition to be represented mathematically, however, the mathematics that has developed to describe logic has, in recent years, been crucial in the design of computer circuits and in automation.

Charles L Dodgson, 1832 -1898, (who under the pseudonym *Lewis Carroll* wrote 'Alice in Wonderland') was an Oxford mathematician who wrote about logic. One example of his logic problems concerns Mrs Bond's ducks.

Activity 1   Do ducks wear collars?

The following lines are taken from Lewis Carroll's book 'Symbolic Logic' first published in 1897.

>"All ducks in this village, that are branded 'B' belong to Mrs Bond;

>Ducks in this village never wear lace collars, unless they are branded 'B';

>Mrs Bond has no grey ducks in this village."

Is the conclusion 'no grey ducks in this village wear lace collars' valid?

8.1   The nature of logic

The Greek philosopher *Aristotle* (384 - 322 BC) is considered to be
the first to have studied logic in that he formed a way of
representing **logical propositions** leading to a conclusion.
Aristotle's theory of **syllogisms** provides a way of analysing
propositions given in the form of statements.

For example, here are some propositions.

> All apples are fruits.

> No toothache is pleasant.

> Some children like chocolate.

> Some cheese is not pasteurised.

In each of these statements a **subject**, S (e.g. apples) is linked to a
predicate, P (e.g. fruits). The **quantity** of each subject is indicated
by the word 'all', 'no' or 'some'.

Statements can also be described as **universal** ('all' or 'no') or
particular ('some') and **affirmative** or **negative**.

The four statements above can be described in this way:

all S is P	universal affirmative
no S is P	universal negative
some S is P	particular affirmative
some S is not P	particular negative .

Aristotle described an argument by linking together three
statements; two statements, called **premises,** lead to a third
statement which is the **conclusion** based on the premises. This
way of representing an **argument** is called a **syllogism.**

Example

> If fruits are tasty
>
> and apples are fruits
> ___
> then apples are tasty.

In this example of a syllogism the conclusion of the argument has
apples as subject (S) and tasty as predicate (P). The first premise
includes P and the second premise includes S and both the first two
premises include 'fruit', which is known as the middle term (M).

The syllogism can therefore be described as:

> If M is P
>
> and S is M
> _____
>
> then S is P.

Activity 2 Finding the figure

By removing all the words the last example can be described as

> M P
>
> S M
> ___
> S P

On the assumption that

- the conclusion of the argument must be SP,
- the first premise must contain P,
- the second premise must contain S,
- both the first and the second premise must contain M,

find the three other arrangements.

Together these are known as the **four figures** of the syllogism.

Not all syllogisms are valid

Each of the four figures can be universal or particular, affirmative or negative, but not all these arrangements give valid arguments.

Example

Is this valid?

> No M is P
>
> All S is M
> _____
> Some S is P.

Solution

An example of this syllogism might be:

No animals with 4 legs is a bird

All cats are animals with 4 legs

Some cats are birds.

Obviously this arrangement is an invalid argument!

Activity 3   Valid or invalid?

Decide if these syllogisms are valid.

(a) Some M is P

All   S is M

All   S is P

(b) All   P is M

No   S is M

No   S is P

(c) No   M is P

All   M is S

Some S is P

(d) Some P is M

All   M is S

Some S is P

8.2   Combining propositions

Modern logic is often called **propositional logic**; the word
'proposition' is defined as a statement that is either **true** or **false**.
So far, a variety of propositions have been considered, such as
premises and conclusions to an argument.

For example, consider the statement

'the water is deep'.

It is not possible to say if this is true or false unless the word 'deep'
is defined and, without a precise definition, this cannot be called a
proposition.

Example

- **p** stands for the proposition 'January has 31 days',
   which is true.

- **q** stands for the proposition '$4 + 7 = 10$',
   which is false.

- 'What a hot day' is not a proposition because it is not in
   subject-predicate form; also the word 'hot' is not defined.

- 'What a hot day' is not a proposition because it is not in
   subject-predicate form; also the word 'hot' is not defined.

Negation NOT ~

Each proposition has a corresponding negation and, if the proposition is denoted by **p**, the negation of the proposition is denoted by ~**p**, read as 'not **p**'.

Example

If **p** is the proposition 'the table is made of pine',

then ~**p** is the proposition 'the table is not made of pine'.

If **q** is the proposition 'the sack is empty', then ~ **q** is the statement 'the sack is not empty'. It is not correct to assume that the negation is 'the sack is full', since the statement 'the sack is not empty' could mean 'the sack is only partly full'.

Connectives

Simple propositions such as

'Elgar composed the Enigma Variations'

'Elgar lived in Malvern'

can be joined by the **connective** 'and' to form a **compound proposition** such as

'Elgar composed the Enigma Variations **and** lived in Malvern.'

A **compound proposition** can be described as a proposition made up of two or more simple propositions joined by connectives.

There is a variety of connectives which will now be defined.

Conjunction AND ∧

If two propositions are joined by the word AND to form a compound statement, this is called a **conjunction** and is denoted by the symbol ∧.

Example

If **p** is the proposition 'the sun is shining'

and **q** is the proposition 'Jack is wearing sunglasses',

then **p** ∧**q** represents the conjunction 'the sun is shining AND Jack is wearing sunglasses'

Note

In the printed form, a proposition is shown by a small letter printed bold, eg **a**, **b**.

When writing by hand it is not easy to make letters bold, so each letter is written with a squiggle underneath it (as with vectors), to indicate that it is a proposition.

For example, the proposition printed as **p** would be hand written as p.
  ~

Disjunction OR ∨

If two statements are joined by the word OR to form a compound proposition, this is called a **disjunction** and is denoted by the symbol ∨.

Example

If **p** is the proposition 'Ann is studying geography'

and **q** is the proposition 'Ann is studying French'

then the disjunction **p** ∨**q** is the compound statement

'Ann is studying geography OR French.'

The word 'OR' in this context can have two possible meanings.

Can Ann study both subjects?

Think about the meaning of these two sentences.

'I can deliver your coal on Wednesday or Thursday.'

'My fire can burn logs or coal.'

The first sentence implies that there is only one delivery of coal and illustrates the **exclusive** use of OR, meaning 'or' but not 'both'. The coal can be delivered on Wednesday or Thursday, but would not be delivered on both days.

The second sentence illustrates the **inclusive** use of 'OR' meaning that the fire can burn either logs or coal, or both logs and coal.

The word 'OR' and the symbol '∨' are used for the **inclusive** OR, which stands for 'and/or'.

The exclusive OR is represented by the symbol ⊕.

Activity 4   Exclusive or inclusive?

Write down three English sentences which use the inclusive OR and three which use the exclusive OR.

Now that a range of connectives is available propositions can be combined into a variety of compound propositions.

Example

Use **p**, **q** and **r** to represent affirmitive (or positive) statements and express the following proposition symbolically.

'Portfolios may include paintings or photographs but not collages.'

Solution

So, let **p** be 'portfolios may include paintings'

let **q** be 'portfolios may include photographs'

and let **r** be 'portfolios may include collages'.

The proposition therefore becomes

$$(\mathbf{p} \vee \mathbf{q}) \wedge \sim \mathbf{r}.$$

Exercise 8A

1. For each of these compound propositions, use **p**, **q** and **r** to represent affirmative (or positive) statements and then express the proposition symbolically.

   (a) This mountain is high and I am out of breath.

   (b) It was neither wet nor warm yesterday.

   (c) During this school year Ann will study two or three subjects.

   (d) It is not true that $3+7=9$ and $4+4=8$.

2. Let **p** be 'the cooker is working', **q** 'the food supply is adequate' and **r** 'the visitors are hungry'. Write the following propositions in 'plain English':

   (a) $\mathbf{p} \wedge \sim \mathbf{r}$

   (b) $\mathbf{q} \wedge \mathbf{r} \wedge \sim \mathbf{p}$

   (c) $\mathbf{r} \vee \sim \mathbf{q}$

   (d) $\sim \mathbf{r} \vee (\mathbf{p} \wedge \mathbf{q})$

   (e) $\sim \mathbf{q} \wedge (\sim \mathbf{p} \vee \sim \mathbf{r})$

8.3   Boolean expressions

The system of logic using expressions such as $\mathbf{p} \vee \mathbf{q}$ and $\sim \mathbf{p} \wedge \mathbf{r}$ was developed by the British mathematician *George Boole* (1815 - 1864).

The laws of reasoning were already well known in his time and Boole was concerned with expressing the laws in terms of a special algebra which makes use of what are known as **Boolean expressions**, such as $\sim \mathbf{a} \wedge \mathbf{b}.$

Activity 5   Using plain English

Define three propositions of your own **p, q** and **r,** and write in plain English the meaning of these Boolean expressions.

  1. $q \wedge r$ 3. $\sim p \vee (q \wedge r)$

  2. $\sim p \wedge r$ 4. $r \wedge (\sim p \vee q)$

Using truth tables

In Section 8.2 a proposition was defined as a statement that is either true or false. In the context of logic, the integers 0 and 1 are used to represent these two states.

    0 represents false

    1 represents true.

Clearly, if a proposition **p** is true then ~**p** is false; also if **p** is false, then ~ **p** is true. This can be shown in a **truth table**, as below.

p	~ p
0	1
1	0

The connectives, $\vee$ and $\wedge$ can also be defined by truth tables, as shown below.

p	q	p∧q
0	0	0
0	1	0
1	0	0
1	1	1

This truth table shows the truth values (0 or 1) of the conjunction $p \wedge q$.

Since $p \wedge q$ means **p** AND **q,** then $p \wedge q$ can only be true (ie 1) when **p** is true AND **q** is true.

If a negation is used, it is best to add the negation column, eg ~**p**, to the truth table.

Example

Construct the truth table for $p \wedge \sim q$.

Solution

p	q	$\sim q$	$p \wedge \sim q$
0	0	1	0
0	1	0	0
1	0	1	1
1	1	0	0

If **p** AND **~q** are true (ie both are 1) then $p \wedge \sim q$ is true.

Exercise 8B

Construct truth tables for the following.

1. $q \vee r$
2. $\sim p \wedge r$
3. $p \vee \sim r$
4. $\sim p \vee \sim q$

8.4 Compound propositions

More complicated propositions can be represented by truth tables, building up parts of the expression.

Example

Construst the truth table for the compound proposition $(a \vee b) \vee \sim c$.

Solution

a	b	c	$(a \vee b)$	$\sim c$	$(a \vee b) \vee \sim c$
0	0	0	0	1	1
0	0	1	0	0	0
0	1	0	1	1	1
0	1	1	1	0	1
1	0	0	1	1	1
1	0	1	1	0	1
1	1	0	1	1	1
1	1	1	1	0	1

Exercise 8C

Construct truth tables for the following:

1. $(a \lor b) \lor c$

2. $a \land (b \land c)$

3. $a \lor (b \lor c)$

4. $(a \land b) \land c$

5. $a \land (b \lor c)$

6. $(a \land b) \lor (a \land c)$

7. $a \lor (b \land c)$

8. $(a \lor b) \land (a \lor c)$

You should notice that some of your answers in this exercise are the same. What are the implications of this? Can you think of similar rules for numbers in ordinary algebra? What names are given to these properties?

8.5 What are the implications?

'If I win this race, then I will be in the finals.'

'If the light is red, then you must stop.'

These two sentences show another connective, 'if ... then ...' which is indicated by the symbol $\Rightarrow$.

$x \Rightarrow y$ is the compound proposition meaning that proposition x implies proposition y.

Returning to the compound proposition

'If I win this race, then I will be in the finals',

this can be written as $a \Rightarrow b$ where a is the proposition 'I win the race' and b is the proposition 'I will be in the finals'. The first proposition, a, 'I win this race', can be true or false. Likewise, the second proposition 'I will be in the finals' can be true or false.

If I win the race (a is true) and I am in the final (b is true) then the compound proposition is true ($a \Rightarrow b$ is true).

Activity 6   The implication truth table

If I fail to win the race (a is false) and I am not in the final (b is false), is the compound proposition $a \Rightarrow b$ true or false?

If I win the race but am not in the final (illness, injury), then is the compound proposition $a \Rightarrow b$ true or false?

By considering these two questions and two others, it is possible to build up a truth table for the proposition $a \Rightarrow b$. Think about the other two questions and their answers, and hence complete the following truth table.

a	b	a ⇒ b
0	0	
0	1	
1	0	
1	1	

The values in this truth table often cause much argument, until it is realised that the connective ⇒ is about implication and not about cause and effect.

It is not correct to assume that **a** ⇒ **b** means **a** causes **b** or that **b** results from **a**.

In fact, the implication connective, ⇒, is defined by the values shown in the truth table whatever the propositions that make up the compound proposition.

Consider the implication **a** ⇒ **b**

'If it is hot, it is June.'

The only way of being sure that this implication **a** ⇒ **b** is false is by finding a time when it is hot but it isn't June; i.e. when **a** is true but **b** is false. Hence the truth table for **a** ⇒ **b** is as follows:

a	b	a ⇒ b
0	0	1
0	1	1
1	0	0
1	1	1

In logic, the two propositions which make up a compound proposition may not be related in the usual sense.

Example

'If Christmas is coming (C), today is Sunday (S).'

C	S	C $\Rightarrow$ S
0	0	1
0	1	1
1	0	0
1	1	1

However difficult it may seem to invent a meaning for this implication, the truth table will be exactly the same as before.

Exercise 8D

1. Give the truth values (1 or 0) of these propositions.

    (a) If all multiples of 9 are odd, then multiples of 3 are even.

    (b) If dogs have four legs then cats have four legs.

    (c) If the sea is blue, the sky is green.

    (d) Oxford is in Cornwall if Sheffield is in Yorkshire.

    (e) Pentagons have six sides implies that quadrilaterals have four sides.

2. If **a** represents 'the crops grow', **b** is 'I water the plants' and **c** is 'I spread manure', express these propositions in terms of **a**, **b** and **c**.

    (a) If I water the plants the crops grow.

    (b) I do not spread manure nor do I water the plants and the crops do not grow.

    (c) If I spread manure the crops grow.

    (d) The crops grow if I water the plants and do not spread manure.

    (e) If I do not water the plants, then I spread manure and the crops grow.

3. Using **a**, **b** and **c** from Question 2, interpret the following propositions.

    (a) $(\mathbf{a} \vee \mathbf{b}) \wedge (\mathbf{a} \vee \mathbf{c})$

    (b) $(\mathbf{c} \vee \sim \mathbf{b}) \Rightarrow \sim \mathbf{a}$

    (c) $\mathbf{a} \Rightarrow \mathbf{b} \wedge \mathbf{c}$

    (d) $\sim \mathbf{a} \vee \mathbf{c} \Rightarrow \mathbf{b}$

8.6   Recognising equivalence

There is a difference between the proposition

'If it is dry, I will paint the door.'

and the proposition

'If, and only if, it is dry, I will paint the door.'

If **p** is 'it is dry' and **q** is 'I will paint the door', then $\mathbf{p} \Rightarrow \mathbf{q}$ represents the first proposition.

The second proposition uses the connective of **equivalence** meaning 'if and only if ' and is represented by the symbol ⇔, i.e. **p** ⇔ **q** represents the second proposition.

The truth table for **p** ⇔ **q** shown here is more obvious than the truth table for implication.

p	q	p⇔q
0	0	1
0	1	0
1	0	0
1	1	1

You will see that **p** ⇔ **q** simply means that the two propositions **p** and **q** are true or false together: this accounts for the use of the word 'equivalence'. Note that if you work out the truth table for **q** ⇔ **p** you will get the same results as for **p** ⇔ **q**.

Exercise 8E

1. If **a** is a true statement and **b** is false, write down the truth value of:

   (a) $a \Leftrightarrow \sim b$

   (b) $\sim b \Leftrightarrow \sim a$.

2. If **a** is 'the theme park has excellent rides', **b** is 'entrance charges are high' and **c** is 'attendances are large', write in plain English the meaning of:

   (a) $c \Leftrightarrow (a \wedge \sim b)$

   (b) $(\sim c \vee \sim b) \Rightarrow a$.

8.7 Tautologies and contradictions

In the field of logic, a **tautology** is defined as a compound proposition which is **always true** whatever the truth values of the constituent statements.

p	~p	p∨~p
0	1	1
1	0	1

This simple truth table shows that

   p∨ ~ **p** is a tautology.

The opposite of a tautology, called a **contradiction**, is defined as a compound proposition which is **always false** whatever the truth values of the constituent statements.

p	~p	p∧~p
0	1	0
1	0	0

This simple truth table shows that

$p \land \sim p$ is a contradiction.

Example

Is $[a \land (b \lor \sim b)] \Leftrightarrow a$ a tautology or a contradiction?

Solution

The clearest way to find the solution is to draw up a truth table. If the result is always true then the statement is a tautology; if always false then it is a contradiction.

a	b	~b	(b∨~b)	a∧(b∨~b)	[a∧(b∨~b)]⇔a
0	0	1	1	0	1
0	1	0	1	0	1
1	0	1	1	1	1
1	1	0	1	1	1

The truth table shows that, since the compound statement is always true, the example given is a tautology.

Exercise 8F

Decide whether each of the following is a tautology or a contradiction.

1. $(a \Rightarrow b) \Leftrightarrow (a \land \sim b)$

2. $[a \land (a \Rightarrow b)] \land \sim b$

3. $\sim(a \Rightarrow b) \Rightarrow [(b \lor c) \Rightarrow (a \lor c)]$

8.8 The validity of an argument

In Section 8.1 the idea of an argument was described as a set of premises (such as **p**, **q** and **r**) which leads to a conclusion (**c**):

$$\mathbf{p}$$
$$\mathbf{q}$$
$$\mathbf{r}$$
$$\cdot$$
$$\cdot$$
$$\cdot$$
$$\overline{\mathbf{c}}$$

A **valid** argument is one in which, if the premises are true, the conclusion must be true. An **invalid** argument is one that is not valid. The validity of an argument can, in fact, be independent of the truth (or falsehood) of the premises. It is possible to have a valid argument with a false conclusion or an invalid argument with a true conclusion. An argument can be shown to be valid if $\mathbf{p} \wedge \mathbf{q} \wedge \mathbf{r} \wedge .... \Rightarrow \mathbf{c}$ is always true (i.e. a tautology).

Example

Represent the following argument symbolically and determine whether the argument is valid.

> If cats are green then I will eat my hat.
>
> I will eat my hat.
> _____
> Cats are green.

Solution

Write the argument as

$$\mathbf{a} \Rightarrow \mathbf{b}$$
$$\underline{\mathbf{b}}$$
$$\mathbf{a}$$

The argument is valid if $(\mathbf{a} \Rightarrow \mathbf{b}) \wedge \mathbf{b} \Rightarrow \mathbf{a}$.

a	b	$(\mathbf{a} \Rightarrow \mathbf{b})$	$(\mathbf{a} \Rightarrow \mathbf{b}) \wedge \mathbf{b}$	$(\mathbf{a} \Rightarrow \mathbf{b}) \wedge \mathbf{b} \Rightarrow \mathbf{a}$
0	0	1	0	1
0	1	1	1	0
1	0	0	0	1
1	1	1	1	1

The truth table shows that the argument is not always true (i.e. it is not a tautology) and is therefore invalid. The second line in the truth table shows that the two premises $\mathbf{a} \Rightarrow \mathbf{b}$ and $\mathbf{b}$ can both be true with the conclusion $\mathbf{a}$ being false. In other words, the compound proposition $(\mathbf{a} \Rightarrow \mathbf{b}) \wedge \mathbf{b} \Rightarrow \mathbf{a}$ is not always true (i.e. it is not a tautology). Therefore the argument

$$\mathbf{a} \Rightarrow \mathbf{b}$$
$$\underline{\mathbf{b}}$$
$$\mathbf{a}$$

is invalid.

Exercise 8G

Determine whether these arguments are valid.

1.
$$\mathbf{a} \Rightarrow \mathbf{b}$$
$$\underline{\mathbf{a} \Rightarrow \mathbf{c}}$$
$$\mathbf{a} \Rightarrow (\mathbf{b} \wedge \mathbf{c})$$

2.
$$\sim \mathbf{b} \Rightarrow \sim \mathbf{a}$$
$$\underline{\mathbf{b}}$$
$$\mathbf{a}$$

3.
$$\mathbf{p} \Rightarrow \mathbf{q}$$
$$\underline{\mathbf{r} \Rightarrow \sim \mathbf{q}}$$
$$\mathbf{p} \Rightarrow \sim \mathbf{r}$$

4. Form a symbolic representation of the following argument and determine whether it is valid.

   If I eat well then I get fat.

   If I don't get rich then I don't get fat.

   I get rich.

8.9 Miscellaneous Exercises

1. Denote the positive (affirmative) statements in the following propositions by $\mathbf{a}, \mathbf{b}, \mathbf{c}, \ldots$ and express each proposition symbolically.

   (a) Either you have understood this chapter, or you will not be able to do this question.

   (b) 64 and 164 are perfect squares.

   (c) $-4 > -9$ and $4 > -9$.

   (d) This is neither the right time nor the right place for an argument.

   (e) If the wind is blowing from the east, I will go sailing tomorrow.

   (f) The train standing at platform 5 will not leave unless all the doors are shut.

   (g) The telephone rang twice and there was no reply.

   (h) My friend will go to hospital if his back doesn't get better.

2. Draw up a truth table for these propositions:

   (a) $(\mathbf{p} \vee \sim \mathbf{q}) \Rightarrow \mathbf{q}$

   (b) $[\mathbf{p} \vee (\sim \mathbf{p} \vee \mathbf{q})] \vee (\sim \mathbf{p} \wedge \sim \mathbf{q})$

   (c) $(\sim \mathbf{p} \vee \sim \mathbf{q}) \Rightarrow (\mathbf{p} \wedge \sim \mathbf{q})$

   (d) $\sim \mathbf{p} \Leftrightarrow \mathbf{q}$

   (e) $(\sim \mathbf{p} \wedge \mathbf{q}) \vee (\mathbf{r} \wedge \mathbf{p})$

   (f) $(\mathbf{p} \Leftrightarrow \mathbf{q}) \Rightarrow (\sim \mathbf{p} \wedge \mathbf{q})$

3. Decide whether each of the following is a tautology:

   (a) $\sim \mathbf{a} \Rightarrow (\mathbf{a} \Rightarrow \mathbf{b})$

   (b) $\sim (\mathbf{a} \vee \mathbf{b}) \wedge \mathbf{a}$

   (c) $[\mathbf{a} \wedge (\mathbf{a} \Rightarrow \mathbf{b})] \Rightarrow \mathbf{a}$

   (d) $(\mathbf{a} \Rightarrow \mathbf{b}) \Leftrightarrow \sim (\mathbf{a} \wedge \sim \mathbf{b})$

4. Decide whether each of the following is a contradiction:

    (a) $(a \wedge b) \vee (\sim a \wedge \sim b)$

    (b) $(a \Rightarrow b) \Leftrightarrow (a \wedge \sim b)$

    (c) $\sim (a \wedge b) \vee (a \vee b)$

    (d) $(a \vee b) \Rightarrow \sim (b \vee c)$

5. Formulate these arguments symbolically using **p**, **q** and **r**, and decide whether each is valid.

    (a) If I work hard, then I earn money

    I work hard
    _____

    I earn money

    (b) If I work hard then I earn money

    If I don't earn money then I am not successful
    _____

    I earn money

    (c) I work hard if and only if I am successful

    I am successful
    _____

    I work hard.

    (d) If I work hard or I earn money then I am successful

    I am successful
    _____

    If I don't work hard then I earn money.

*6. *Lewis Carroll* gave many arguments in his book 'Symbolic Logic'. Decide whether the following arguments are valid.

    (a) No misers are unselfish.

    None but misers save egg-shells.
    _____

    No unselfish people save egg-shells.

    (b) His songs never last an hour;

    A song, that lasts an hour, is tedious.
    _____

    His songs are never tedious.

    (c) Babies are illogical;

    Nobody is despised who can manage a crocodile;

    Illogical persons are despised.
    _____

    Babies cannot manage crocodiles.

    (Hint   **a** : persons who are able to manage a crocodile.

            b : persons who are babies

            c : persons who are despised

            d : persons who are logical)

9 BOOLEAN ALGEBRA

Objectives

After studying this chapter you should

- be able to use AND, NOT, OR and NAND gates;
- be able to use combinatorial and switching circuits;
- understand equivalent circuits;
- understand the laws of Boolean algebra;
- be able to simplify Boolean expressions;
- understand Boolean functions;
- be able to minimise circuits;
- understand the significance of half and full adder circuits.

9.0 Introduction

When *George Boole* (1815-1864) developed an algebra for logic, little did he realise that he was forming an algebra that has become ideal for the analysis and design of circuits used in computers, calculators and a host of devices controlled by microelectronics. Boole's algebra is physically manifested in electronic circuits and this chapter sets out to describe the building blocks used in such circuits and the algebra used to describe the logic of the circuits.

9.1 Combinatorial circuits

The circuits and switching arrangement used in electronics are very complex but, although this chapter only deals with simple circuits, the functioning of all microchip circuits is based on the ideas in this chapter. The flow of electrical pulses which represent the **binary digits** 0 and 1 (known as **bits**) is controlled by combinations of electronic devices. These **logic gates** act as switches for the electrical pulses. Special symbols are used to represent each type of logic gate.

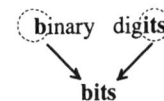

binary digits

bits

NOT gate

The NOT gate is capable of reversing the input pulse. The truth table for a NOT gate is as follows:

Input a	Output ~a
0	1
1	0

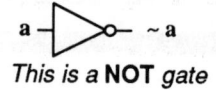

*This is a **NOT** gate*

The NOT gate receives an input, either a pulse (1) or no pulse (0) and produces an output as follows :

If input **a** is 1, output is 0;

and if input **a** is 0, output is 1.

AND gate

The AND gate receives two inputs **a** and **b**, and produces an output denoted by $a \wedge b$. The truth table for an AND gate is as follows :

Input a	b	Output $a \wedge b$
0	0	0
0	1	0
1	0	0
1	1	1

*This is an **AND** gate*

The only way that the output can be 1 is when **a** AND **b** are both 1. In other words there needs to be an electrical pulse at **a** AND **b** before the AND gate will output an electrical pulse.

OR gate

The OR gate receives two inputs **a** and **b**, and produces an output denoted by $a \vee b$. The truth table for an OR gate is as follows:

Input a	b	Output $a \vee b$
0	0	0
0	1	1
1	0	1
1	1	1

*This is an **OR** gate*

The output will be 1 when **a** or **b** or both are 1.

These three gates, NOT, AND and OR, can be joined together to form **combinatorial circuits** to represent Boolean expressions, as explained in the previous chapter.

Example

Use logic gates to represent

(a) $\sim p \vee q$

(b) $(x \vee y) \wedge \sim x$

Draw up the truth table for each circuit

Solution

(a)

p	q	~p	~p ∨ q
0	0	1	1
0	1	1	1
1	0	0	0
1	1	0	1

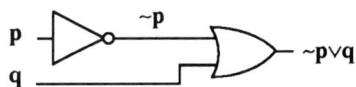

(b)

x	y	x∨y	~x	(x∨y)∧~x
0	0	0	1	0
0	1	1	1	1
1	0	1	0	0
1	1	1	0	0

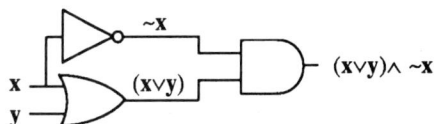

Exercise 9A

Use logic gates to represent these expressions and draw up the corresponding truth tables.

1. $x \wedge (\sim y \vee x)$

2. $a \vee (\sim b \wedge c)$

3. $[a \vee (\sim b \vee c)] \wedge \sim b$

Write down the Boolean expression and the truth table for each of the circuits below.

4.

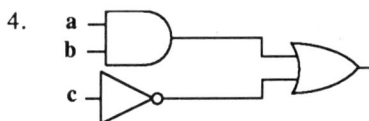

5.

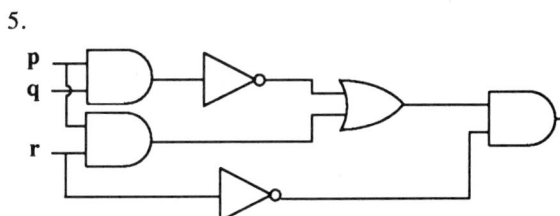

9.2 When are circuits equivalent?

Two circuits are said to be **equivalent** if each produce the same outputs when they receive the same inputs.

Example

Are these two combinatorial circuits equivalent?

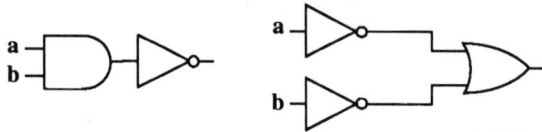

Solution

The truth tables for both circuits will show if they are equivalent :

a	b	$\sim(a \wedge b)$
0	0	1
0	1	1
1	0	1
1	1	0

a	b	$\sim a \vee \sim b$
0	0	1
0	1	1
1	0	1
1	1	0

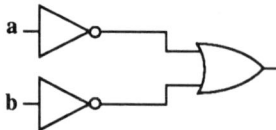

Work through the values in the truth tables for yourself. Since both tables give the same results the two circuits are equivalent. Indeed the two Boolean expressions are equivalent and can be put equal;

i.e. $\sim(a \wedge b) = \sim a \vee \sim b$

Exercise 9B

Show if these combinatorial circuits are equivalent by working out the Boolean expression and the truth table for each circuit.

1.

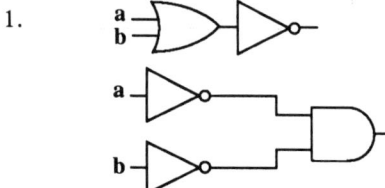

2.

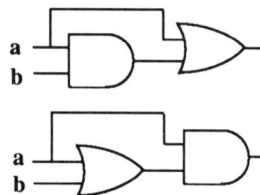

3.

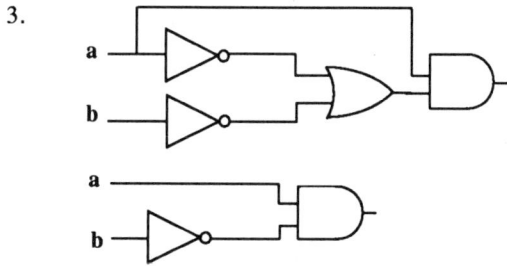

4.

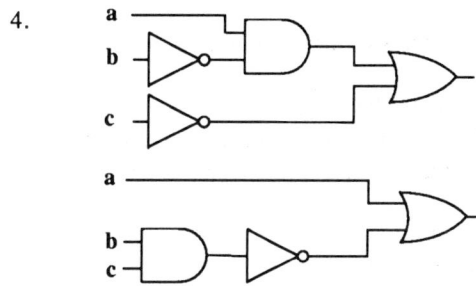

9.3 Switching circuits

A network of switches can be used to represent a Boolean expression and an associated truth table.

Generally the switches are used to control the flow of an electrical current but you might find it easier to consider a switching circuit as a series of water pipes with taps or valves at certain points.

One of the reasons for using switching circuits rather than logic gates is that designers need to move from a combinatorial circuit (used for working out the logic) towards a design which the manufacturer can use for the construction of the electronic circuits.

This diagram shows switches **A**, **B** and **C** which can be **open** or **closed**. If a switch is closed it is shown as a 1 in the following table whilst 0 shows that the switch is open. The **switching table** for this circuit is as follows :

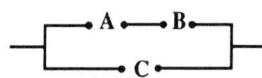

A	B	C	Circuit output
0	0	0	0
0	0	1	1
0	1	0	0
0	1	1	1
1	0	0	0
1	0	1	1
1	1	0	1
1	1	1	1

The table shows that there will be an output (ie 1) when **A** AND **B** are 1 OR **C** is 1. This circuit can therefore be represented as

(A AND B) OR C

ie. $(A \wedge B) \vee C$

The circuit just considered is built up of two fundamental circuits:

• a **series circuit**, often called an AND gate, $A \wedge B$

- **a parallel circuit**, often called an OR gate, **A** ∨**B**.

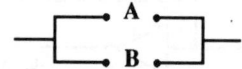

The next step is to devise a way of representing negation. The negation of the truth value 1 is 0 and vice versa, and in switching circuits the negation of a 'closed' path is an 'open' path.

This circuit will always be 'open' whatever the state of **A**. In other words the output will always be 0, irrespective of whether **A** is 1 or 0.

This circuit will always be 'closed' whatever the state of **A**. The output will always be 1 irrespective of whether **A** is 1 or 0.

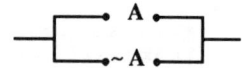

Example

Represent the circuit shown opposite symbolically and give the switching table.

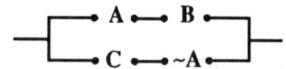

Solution

The symbolic representation can be built up by considering

the top line of the circuit   (**A** ∧**B**)

the top bottom of the circuit   (**C**∧~**A**).

Combining these gives the result (**A** ∧**B**)∨ (**C**∧~**A**)

The table is as follows.

A	**B**	**C**	~**A**	**A**∧**B**	**C**∧~**A**	(**A**∧**B**)∨(**C**∧~**A**)
0	0	0	1	0	0	0
0	0	1	1	0	1	1
0	1	0	1	0	0	0
0	1	1	1	0	1	1
1	0	0	0	0	0	0
1	0	1	0	0	0	0
1	1	0	0	1	0	1
1	1	1	0	1	0	1

Activity 1   Make your own circuit

Using a battery, some wire, a bulb and some switches, construct the following circuit.

A simple switch can be made using two drawing pins and a paper clip which can swivel to close the switch. The pins can be pushed into a piece of corrugated cardboard or polystyrene.

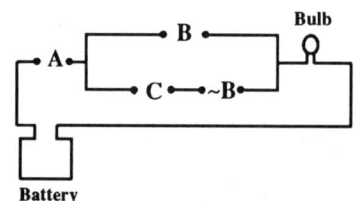

Using the usual notation of 1 representing a closed switch and 0 representing an open switch, you can set the switches to represent each line of this table:

A	B	C	Output
0	0	0	
0	0	1	
0	1	0	
0	1	1	
1	0	0	
1	0	1	
1	1	0	
1	1	1	

Remember that the '~B' switch is always in the opposite state to the 'B' switch.

Record the output using 1 if the bulb lights up (i.e. circuit is closed) and 0 if the bulb fails to light (i.e. circuit is open)

Represent the circuit symbolically and draw up another table to see if you have the same output.

Exercise 9C

Represent the following circuits by Boolean expresions:

1.

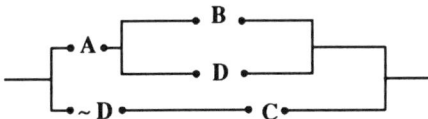

2.

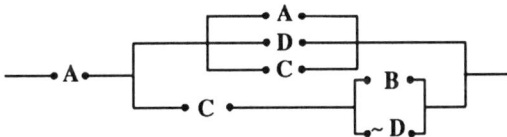

Draw switching circuits for these Boolean expressions

3. $A \vee (\sim B \wedge C)$

4. $A \wedge ((\sim B \wedge C) \vee (B \wedge \sim C))$

9.4   Boolean algebra

A variety of Boolean expressions have been used but *George Boole* was responsible for the development of a complete algebra. In other words, the expressions follow laws similar to those of the algebra of numbers.

The operators $\wedge$ and $\vee$ have certain properties similar to those of the arithmetic operators such as $+$, $-$, $\times$ and $\div$.

(a)   **Associative laws**

$$(a \vee b) \vee c = a \vee (b \vee c)$$

and $$(a \wedge b) \wedge c = a \wedge (b \wedge c)$$

(b)   **Commutative laws**

$$a \vee b = b \vee a$$

and $$a \wedge b = b \wedge a$$

(c)   **Distributive laws**

$$a \wedge (b \vee c) = (a \wedge b) \vee (a \wedge c)$$

and $$a \vee (b \wedge c) = (a \vee b) \wedge (a \vee c)$$

These laws enable Boolean expressions to be simplified and another law developed by an Englishman, *Augustus de Morgan* (1806-1871) is useful. He was a contemporary of *Boole* and worked in the field of logic and is now known for one important result bearing his name:-

(d)   **de Morgan's laws**

$$\sim(a \vee b) = \; \sim a \wedge \sim b$$

and $$\sim(a \wedge b) = \; \sim a \vee \sim b$$

Note: You have to remember to change the connection,
$\wedge$ changes to $\wedge$, $\vee$ changes to $\wedge$.

Two more laws complete the range of laws which are included in the Boolean algebra.

(e)   **Identity laws**

$$a \vee 0 = a$$

open switch $= 0$

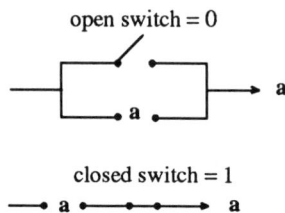

and $$a \wedge 1 = a$$

closed switch $= 1$

(f) Complement laws

$$a \lor \sim a = 1$$

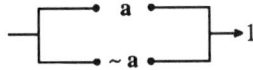

$$\text{and} \quad a \land \sim a = 0$$

The commutative law can be developed to give a further result which is useful for the simplification of circuits.

Consider the expressions $a \land (a \lor b)$ and the corresponding circuit.

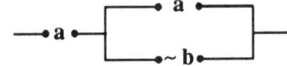

If switch a is open $(a = 0)$ **what can you say about the whole circuit? What happens when switch a is closed** $(a = 1)$**? Does the switch b have any effect on your answers?**

The truth table for the circuit above shows that $a \land (a \lor b) = a$.

a	b	$a \land (a \lor b)$
0	0	0
0	1	0
1	0	1
1	1	1

This result can be extended to more switches. For example

if $\qquad a \land (a \lor b) = a$

then $\qquad a \land (a \lor b \lor c) = a$

and $\qquad a \land (a \lor b \lor c) \land (a \lor b) \land (b \lor c) = a \land (b \lor c).$

The last of these expressions is represented by this circuit:

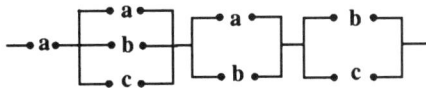

which can be replaced by the simplified circuit:

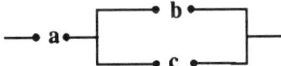

Example

Write down a Boolean expression for this circuit. Simplify the expression and draw the corresponding circuit.

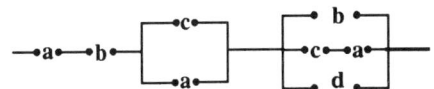

Solution

$$\underbrace{a \wedge b \wedge (a \vee c)} \wedge \left(b \vee (c \wedge a) \vee d\right)$$
$$a \wedge (a \vee c) \wedge b \wedge \left(b \vee (c \wedge a) \vee d\right)$$

Since $a \wedge (a \vee c) = a$

and $b \wedge \left(b \vee (c \wedge a) \vee d\right) = b,$

an equivalent expression is $a \wedge b$

and the circuit simplifies to •———• a •———• b •———•

Activity 2   Checking with truth tables

Draw truth tables for the example above to check that

$$a \wedge b \wedge (a \vee c) \wedge \left(b \vee (c \wedge a) \vee d\right) = a \wedge b$$

Exercise 9D

Simplify the following and check your answers by drawing up truth tables:

1. $a \vee (\sim a \wedge b)$

2. $a \wedge \left[b \vee (a \wedge b)\right] \wedge \left[a \vee (\sim a \wedge b)\right]$

3. Simplify the following circuit:

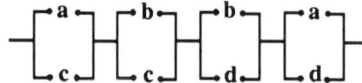

9.5   Boolean functions

In the same way as algebraic functions describe the relationship between the domain, (a set of inputs) and the range (a set of outputs), a **Boolean function** can be described by a **Boolean expression**. For example, if

$$f\left(x_1, x_2, x_3\right) = x_1 \wedge \left(\sim x_2 \vee x_3\right)$$

then f is the Boolean function and

$$x_1 \wedge \left(\sim x_2 \vee x_3\right)$$

is the Boolean expression.

Example

Draw the truth table for the Boolean function defined as

$$f\left(x_1, x_2, x_3\right) = x_1 \wedge \left(\sim x_2 \vee x_3\right)$$

Solution

The inputs and outputs of this Boolean function are shown in the following table:

x_1	x_2	x_3	$f\left(x_1, x_2, x_3\right)$
0	0	0	0
0	0	1	0
0	1	0	0
0	1	1	0
1	0	0	1
1	0	1	1
1	1	0	0
1	1	1	1

It is sometimes necessary to form a function from a given truth table. The method of achieving this is described in the following example.

Example

For the given truth table, form a Boolean function

a	b	c	f(a, b, c)
0	0	0	1
0	0	1	1
0	1	0	0
0	1	1	0
1	0	0	1
1	0	1	0
1	1	0	1
1	1	1	1

Solution

The first stage is to look for the places where f(**a**,**b**,**c**) is 1 and then link them all together with 'OR's. For example, in the last row f(**a**, **b**, **c**) = 1 and this is the row in which **a**, **b** and **c** are all true; i.e. when $a \wedge b \wedge c = 1$.

The output is also 1, (i.e. $f(\mathbf{a}, \mathbf{b}, \mathbf{c}) = 1$) in the 7th row which leads
to the combination

$$\mathbf{a} \wedge \mathbf{b} \wedge \sim \mathbf{c} = 1$$

Similarly, for the 5th row

$$\mathbf{a} \wedge \sim \mathbf{b} \wedge \sim \mathbf{c} = 1$$

and for the 2nd row

$$\sim \mathbf{a} \wedge \sim \mathbf{b} \wedge \mathbf{c} = 1$$

and for the 1st row

$$\sim \mathbf{a} \wedge \sim \mathbf{b} \wedge \sim \mathbf{c} = 1$$

All these combinations are joined using the connective $\vee$ to give
the Boolean expression

$$(\mathbf{a} \wedge \mathbf{b} \wedge \mathbf{c}) \vee (\mathbf{a} \wedge \mathbf{b} \wedge \sim \mathbf{c}) \vee (\mathbf{a} \wedge \sim \mathbf{b} \wedge \sim \mathbf{c}) \vee (\sim \mathbf{a} \wedge \sim \mathbf{b} \wedge \mathbf{c}) \vee (\sim \mathbf{a} \wedge \sim \mathbf{b} \wedge \sim \mathbf{c})$$

If the values of $\mathbf{a}$, $\mathbf{b}$ and $\mathbf{c}$ are as shown in the 1st, 2nd, 5th, 7th and
8th row then the value of $f(\mathbf{a}, \mathbf{b}, \mathbf{c}) = 1$ in each case, and the
expression above has a value of 1. Similarly if $\mathbf{a}, \mathbf{b}$ and $\mathbf{c}$ are as
shown in the table for which $f(\mathbf{a}, \mathbf{b}, \mathbf{c}) = 0$ then the expression
above has the value of 0.

The Boolean function for the truth table is therefore given by

$$f(\mathbf{a}, \mathbf{b}, \mathbf{c}) = (\mathbf{a} \wedge \mathbf{b} \wedge \mathbf{c}) \vee (\mathbf{a} \wedge \mathbf{b} \wedge \sim \mathbf{c}) \vee (\mathbf{a} \wedge \sim \mathbf{b} \wedge \sim \mathbf{c}) \vee (\sim \mathbf{a} \wedge \sim \mathbf{b} \wedge \mathbf{c}) \vee (\sim \mathbf{a} \wedge \sim \mathbf{b} \wedge \sim \mathbf{c})$$

This is called the **disjunctive normal form** of the function f; the
combinations formed by considering the rows with an output value
of 1 are joined by the disjunctive connective, OR.

Exercise 9E

Find the disjunctive normal form of the Boolean
function for these truth tables:

1.

a	b	f (a, b)
0	0	1
0	1	0
1	0	1
1	1	0

2.

a	b	f(a, b)
0	0	1
0	1	1
1	0	0
1	1	1

3.

x	y	z	f (x, y, z)
0	0	0	1
0	0	1	0
0	1	0	0
0	1	1	0
1	0	0	1
1	0	1	0
1	1	0	0
1	1	1	1

9.6   Minimisation with NAND gates

When designing combinatorial circuits, efficiency is sought by minimising the number of gates (or switches) in a circuit. Many computer circuits make use of another gate called a NAND gate which is used to replace NOT AND, thereby reducing the number of gates.

The NAND gate receives inputs **a** and **b** and the output is denoted by $a \uparrow b$.

The symbol used is

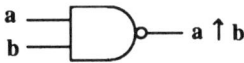

The truth table for this is

a	b	$a \uparrow b$
0	0	1
0	1	1
1	0	1
1	1	0

The NAND gate is equivalent to

Note that, by de Morgan's law, $\sim(a \wedge b) = \sim a \vee \sim b$.

Example

Use NAND gates alone to represent the function

$$f(a,\ b,\ c,\ d) = (a \wedge b) \vee (c \wedge d)$$

Solution

The use of NAND gates implies that there must be negation so the function is rewritten using de Morgan's Laws:

$$(a \wedge b) \vee (c \wedge d) = \ \sim\big[(\sim a \ \vee \sim b) \wedge (\sim c \ \vee \sim d)\big]$$

The circuit consisting of NAND gates is therefore as follows:

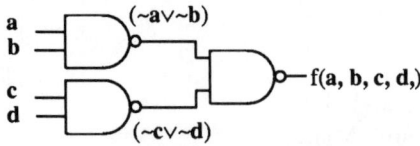

Example

Design combinatorial circuits to represent (a) the negation function $f(x) = \sim x$ and (b) the OR function $f(x, y) = x \vee y$.

Solution

(a) $\sim x = \sim (x \wedge x)$

(b) $x \vee y = \sim (\sim x \vee \sim y)$

   $\qquad = \sim x \uparrow \sim y$

   $\qquad = (x \uparrow x) \uparrow (y \uparrow y)$

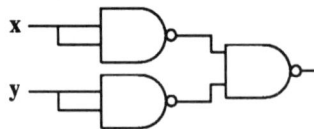

Exercise 9F

Design circuits for each of the following using only NAND gates.

1. $a \wedge b$
2. $a \wedge \sim b$
3. $(\sim a \wedge \sim b) \vee \sim b$

9.7   Full and half adders

Computers turn all forms of data into **binary digits** , (0 s and 1 s), called bits, which are manipulated mathematically. For example the number 7 is represented by the binary code 00000111 (8 bits are used because many computers use binary digits in groups of 8, for example, ASCII code). This section describes how binary digits can be added using a series of logic gates. The basic mathematical operation is **addition** since

> **subtraction**   is the addition of negative numbers,
>
> **multiplication**   is repeated addition,
>
> **division**   is repeated subtraction.

When you are adding two numbers there are two results to note for each column; the entry in the answer and the carrying figure.

$$\begin{array}{r} 254 \\ 178 \\ \hline 432 \\ \hline 11 \end{array}$$

When 4 is added to 8 the result is 12, 2 is noted in the answer and the digit 1 is carried on to the next column.

When adding the second column the carry digit from the first column is included, i.e. $5+7+1$, giving yet another digit to carry on to the next column.

Half adder

The half adder is capable of dealing with two inputs, i.e. it can only add two bits, each bit being either 1 or 0.

a	b	Carry bit	Answer bit
0	0	0	0
0	1	0	1
1	0	0	1
1	1	1	1

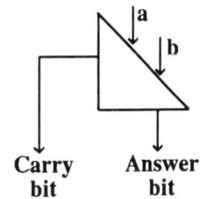

Carry bit   Answer bit

Activity 3   Designing the half adder circuit

The next stage is to design a circuit which will give the results shown in the table above.

The first part of the circuit is shown opposite; complete the rest of the circuit which can be done with a NOT gate, an OR gate and an AND gate to give the answer bit.

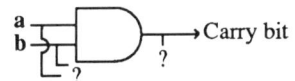

Full Adder

A half adder can only add two bits; a full adder circuit is capable of including the carry bit in the addition and therefore has three inputs.

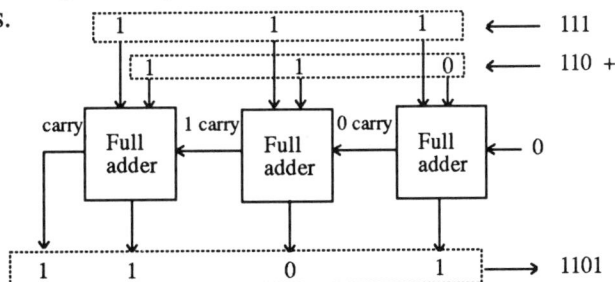

149

Activity 4   Full adder truth table

Complete this truth table for the full adder.

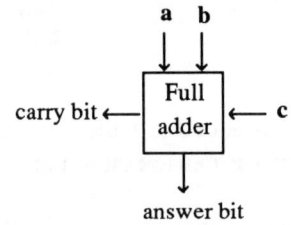

Inputs				
a	b	c	Carry bit	Answer bit
0	0	0	0	0
0	0	1	0	1
0	1	0		
	etc ↓			
1	1	1		

The circuit for a full adder is, in effect, a combination of two half adders.

If you think about it, the carry bit of the full adder must be 1 if either of the two half adders shown gives a carry bit of 1 (and in fact it is impossible for both those half adders to give a carry bit of 1 at the same time). Therefore the two carry bits from the half adders are fed into an OR gate to give an output equal to the carry bit of the full adder.

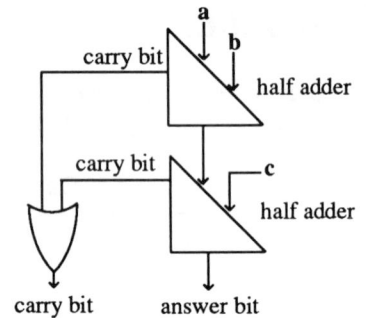

The circuit for a full adder consists, therefore, of two half adders with the carry bits feeding into an OR gate as follows:

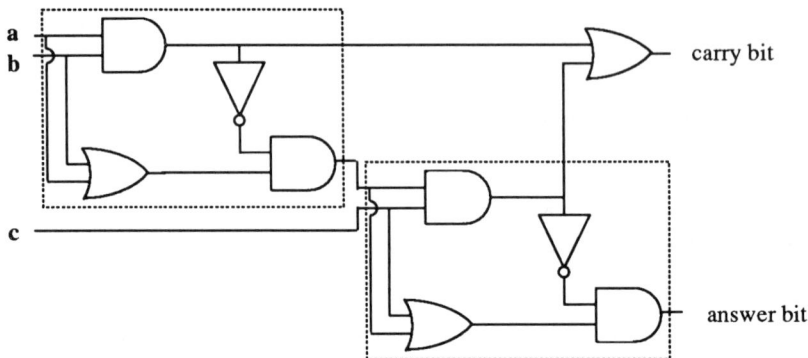

The dotted lines enclose the two half adders with the whole circuit representing a full adder.

Activity 5   NAND half adder

Draw up a circuit to represent a half adder using only NAND gates.

9.8 Miscellaneous Exercises

1. Use logic gates to represent these expressions and draw up the corresponding truth tables:

   (a) $\sim\!\left[(a \wedge b) \vee c\right]$

   (b) $(a \wedge b) \vee \sim\! c$

   (c) $\sim\! c \wedge \left[(a \wedge b) \vee \sim\!(a \wedge c)\right]$

2. Write down the Boolean expression and the truth table for each of these circuits:

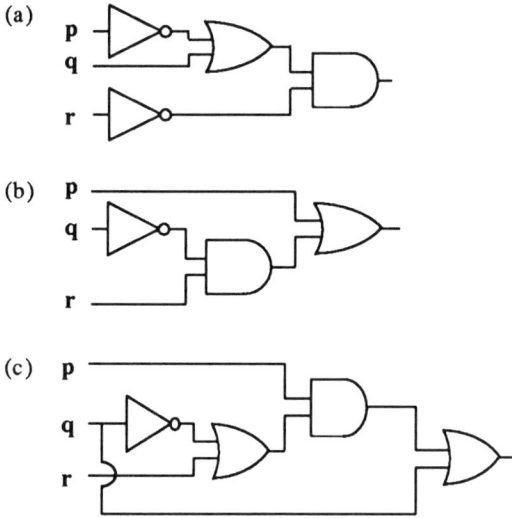

   (a)

   (b)

   (c)

3. Write down the Boolean expressions for these circuits :

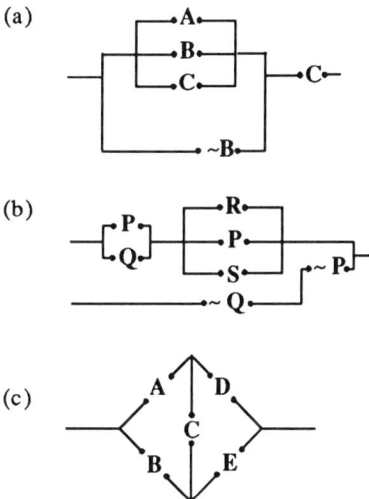

   (a)

   (b)

   (c)

4. Draw switching circuits for these Boolean expressions:

   (a) $(B \wedge C) \vee (C \wedge A) \vee (A \wedge B)$

   (b) $\left[A \wedge \left((B \wedge \sim\! C) \vee (\sim\! B \wedge C)\right)\right] \vee (\sim\! A \wedge B \wedge C)$

5. Draw the simplest switching circuit represented by this table:

P	Q	R	S	Output
1	1	1	1	1
0	1	1	1	1
1	1	0	1	1
0	1	0	1	1

6. A burglar alarm for a house is controlled by a switch. When the switch is on, the alarm sounds if either the front or back doors or both doors are opened. The alarm will not work if the switch is off. Design a circuit of logic gates for the alarm and draw up the corresponding truth table.

*7. A gallery displaying a famous diamond uses a special Security Unit to protect access to the Display Room (D). The diagram below shows the layout of the system.

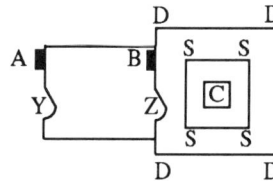

The display cabinet (C) is surrounded by a screen of electronic eyes (S).

Access to the display room is through doors (Y), (Z). Boxes (A), (B) are used in the system. The following persons are involved in the system :

> Manager,
> Deputy Manager,
> Chief Security Officer.

The Display Room is opened as follows :

> The Unit must be activated at box A.

> Door (Y) is opened by any two of the above persons at box A.

> Box B is activated by the Manager and Deputy Manager together.

> The screen (S) is activated by the Chief Security Officer alone at box B only.

> Door (Z) can only be opened once the screen (S) is activated.

Draw a circuit of logic gates required inside the Unit to operate it. Ensure your diagram is documented.

(AEB)

8. Simplify the following expressions and check your answer by drawing up truth tables.

   (a) $(a \wedge b \wedge c) \vee (\sim a \wedge b \wedge c)$

   (b) $a \vee (\sim a \wedge b \wedge c) \vee (\sim a \wedge b \wedge \sim c)$

   (c) $(p \wedge q) \vee (\sim p \vee \sim q) \wedge (r \vee s)$

9. Find the disjunctive normal form of this function; simplify and draw the combinatorial circuit.

a	b	c	f(a,b,c)
0	0	0	0
0	0	1	0
0	1	0	0
0	1	1	1
1	0	0	0
1	0	1	1
1	1	0	1
1	1	1	1

10. Design a circuit representing $\sim a \vee b$ using NAND gates.

*11. Write a computer program or use a spreadsheet that outputs a truth table for a given Boolean expression.

10 DIFFERENCE EQUATIONS I

Objectives

After studying this chapter you should

- be able to detect recursive events within contextual problems;

- be able to recognise and describe associated sequences;

- be able to solve a number of first order difference equations;

- be able to apply solutions of first-order difference equations to contextual problems.

10.0 Introduction

Imagine you are to jump from an aircraft at an altitude of 1000 metres. You want to free-fall for 600 metres, knowing that in successive seconds you fall

$$5, 15, 25, 35, \; ... \text{ metres.}$$

How many seconds do you count before you pull the rip-cord?

Developing a method for answering this type of question is an aim of this chapter. Perhaps you could attempt the problem now by studying the pattern within the sequence.

The methods employed in this chapter are widely used in applied mathematics, especially in areas such as economics, geography and biology. The above example is physical, but as you will see later there is no need to resort to physical or mechanical principles in order to solve the problem.

Activity 1   Tower of Hanoi

You may be familiar with the puzzle called the **Tower of Hanoi**, in which the object is to transfer a pile of rings from one needle to another, one ring at a time, in as few moves as possible, with never a larger ring sitting upon a smaller one.

The puzzle comes from the Far East, where in the temple of Benares a priest unceasingly moves a disc each day from an original pile of sixty four discs on one needle to another. When he has finished the world will end!

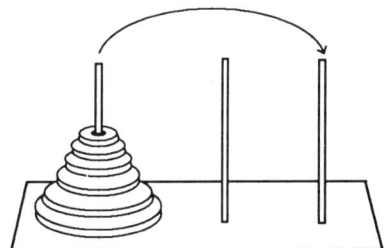

Try this game for yourself. There may be one in school or you could make one. The puzzle is commonly found among mathematics education software.

Record the number of moves required for initial piles of one, two, three rings, etc. Try to predict the number of moves required for 10 rings and 20 rings. When should the world end?

The solution to the problem involves the idea of recursion (from recur - to repeat). The next section considers a further problem through which the ideas of recursion can be explored. You will meet the Tower of Hanoi again, later on.

10.1 Recursion

Here is a simple sequence linked to a triangular dot pattern. Naturally, these are called the triangle numbers 1, 3, 6, 10, 15, ...

In order to obtain the next term (the sixth), one more row of six dots is added to the fifth term. If a term much further down the sequence were required, you could simply keep adding on 7, then 8, then 9 and so on. This process is called **recursion**.

The process can be described algebraically. Call the first term u_1, the second u_2 and the general term u_n, where n is a positive integer.

So $u_1 = 1$

$u_2 = 3$

$u_3 = 6$

etc.

In order to find u_n you have to add the number n to u_{n-1}. This gives the expression

$$u_n = u_{n-1} + n$$

Expressions of this type are called **difference equations** (or **recurrence relations**).

What processes in life are recursive? How, if at all, does natural recursion differ from mathematical recursion?

In order to verify that this expression determines the sequence of triangle numbers, the term u_6 is found from using the known value of u_5 :

$$u_5 = 15$$

and
$$u_6 = u_5 + 6$$
$$= 15 + 6$$
$$= 21, \text{ as expected.}$$

Example

If $u_1 = 4$ and $u_n = 2u_{n-1} + 3n - 1,$ for $n \geq 2,$ find the values of u_2 and u_3.

Solution

$$u_2 = 2u_1 + 3 \times 2 - 1$$
$$= 2 \times 4 + 6 - 1$$
$$= 8 + 6 - 1$$
$$= 13$$

and
$$u_3 = 2u_2 + 3 \times 3 - 1$$
$$= 2 \times 13 + 9 - 1$$
$$= 26 + 9 - 1$$
$$= 34.$$

Activity 2   Dot patterns

Draw a number of dot patterns which increase in a systematic way, for example,

For each pattern, write down a difference equation and show that, from knowing u_1, you can use your equation to generate successive terms in your patterns.

What other patterns occur in life? Can they be described using numbers?

Exercise 10A

1. For each equation you are given the first term of a sequence. Find the 4th term in each case :

    (a) $u_1 = 2$ and $u_n = u_{n-1} + 3$, $n \geq 2$

    (b) $u_1 = 1$ and $u_n = 3u_{n-1} + n$, $n \geq 2$

    (c) $u_1 = 0$ and $u_n - u_{n-1} = n + 1$, $n \geq 2$.

2. For each sequence write down a difference equation which describes it:

    (a) 3 5 7 9 11

    (b) 2 5 11 23 47

    (c) 1 2 5 14 41.

3. A vacuum pump removes one third of the remaining air in a cylinder with each stroke. Form an equation to represent this situation. After how many strokes is just 1 / 1 000 000 of the initial air remaining?

4. Write down the first four terms of each of these sequences and the associated difference equation.

    (a) $u_n = \sum_{r=1}^{n} (2r - 1)$ (b) $u_n = \sum_{r=1}^{n} (10 - r)$

    (c) $u_n = \sum_{r=1}^{n} 3(2r + 1)$

5. Write a simple computer program (say in Basic) which calculates successive terms of a sequence from a difference equation you have met. Here is one for the triangle numbers to help you.

```
10  REM"SEQUENCE OF TRIANGLE NUMBERS"
20  INPUT"NUMBER OF TERMS REQUIRED";N
30  CLS
40  U=0:PRINT"FIRST ";N;" TRIANGLE
    NUMBERS ARE"
50  FOR i=1 TO N:U=U+i:PRINTU:NEXTi
60  STOP
```

10.2 Iteration

Consider again the Tower of Hanoi.

You should have found a sequence of minimum moves as follows :

Number of rings	1	2	3	4	5	6	...
Number of moves	1	3	7	15	31	63	...

Successive terms are easily found by doubling and adding one to the previous term, but it takes quite a long time to reach the sixty-fourth term, which by the way is about 1.85×10^{19} or 18.5 million million million!

A different approach is to try to work 'backwards' from the nth term u_n, rather than starting at u_1, and building up to it. In this case :

$$u_n = 2u_{n-1} + 1, \quad n \geq 2 \qquad (1)$$

For example,

$$u_6 = 2u_5 + 1$$

$$= 2 \times 15 + 1$$

$$= 31.$$

In a similar way, u_{n-1} can be written in terms of u_{n-2} as

$$u_{n-1} = 2u_{n-2} + 1. \qquad (2)$$

Then equation (2) can be substituted into equation (1) to give

$$u_n = 2(2u_{n-2} + 1) + 1$$

$$= 4u_{n-2} + 2 + 1.$$

Repeating this process using $u_{n-2} = 2u_{n-3} + 1$ gives

$$u_n = 2(4u_{n-3} + 2 + 1) + 1$$

$$= 8u_{n-3} + 4 + 2 + 1$$

$$\Rightarrow \quad u_n = 2^3 u_{n-3} + 2^2 + 2^1 + 2^0.$$

You can see a pattern developing. Continuing until u_n is expressed in terms of u_1, gives

$$u_n = 2^{n-1}u_1 + 2^{n-2} + 2^{n-3} + \dots + 2^2 + 2^1 + 2^0$$

$$= 2^{n-1} + 2^{n-2} + \dots + 2^2 + 2^1 + 2^0, \text{ (as } u_1 = 1). \quad (3)$$

You should recognise (3) as a geometric progression (GP) with first term 1 and common ratio 2. Using the formula for the sum to n terms of a GP,

$$u_n = \frac{1(2^n - 1)}{2 - 1} = 2^n - 1, \quad n \geq 1.$$

This is the solution to the differential equation $u_n = 2u_{n-1} + 1$. This process involved the repeated use of a formula and, as in Chapter 3, is known as **iteration**.

You can now see how easy it is to calculate a value for u_n.

For example,

$$u_{100} = 2^{100} - 1$$

$$\approx 127 \times 10^{30}.$$

Activity 3

You invest £500 in a building society for a number of years at a rate of 10% interest per annum. Find out how much will be in the

bank after 1, 2 or 3 years. Try to write down the difference equation which describes the relationship between the amount in the bank, u_n, at the end of the nth year with the amount u_{n-1} at the end of the previous year. Solve your equation by iteration in the way shown for the Tower of Hanoi problem.

Use your solution to find the amount accrued after 10 years.

How long does it take for your money to double?

Exercise 10B

1. Solve by iteration, giving u_n in terms of u_1.

   (a) $u_n = u_{n-1} + 2$, $n \geq 2$

   (b) $u_n = 4u_{n-1} - 1$, $n \geq 2$

   (c) $u_n = 3u_{n-1} + 2$, $n \geq 2$

2. A population is increasing at a rate of 25 per thousand per year. Define a difference equation which describes this situation. Solve it and find the population in 20 years' time, assuming the population is now 500 million. How long will it take the population to reach 750 million?

3. Calculate the monthly repayment on a £500 loan over 2 years at an interest rate of $1\frac{1}{2}\%$ per month.

10.3 First order difference equations

Equations of the type $u_n = ku_{n-1} + c$, where k, c are constants, are called **first order linear difference equations** with constant coefficients. All of the equations you have met so far in this chapter have been of this type, except for the one associated with the triangle numbers in Section 10.1

For the triangle numbers $u_n = u_{n-1} + n$, and since n is not constant, this is not a linear difference equation with constant coefficients.

If the equation is of the type

$$u_n = ku_{n-1},$$

then the solution can be found quite simply. You know that $u_n = ku_{n-1}$, $u_{n-1} = ku_{n-2}$, and so on, so that the difference equation becomes

$$u_n = ku_{n-1}$$

$$= k(ku_{n-2})$$

$$= k^2 u_{n-2}$$

$$= k^2(ku_{n-3})$$

$$= k^3 u_{n-3}$$

etc.

$$\Rightarrow \quad u_n = k^{n-1}u_1.$$

So if $u_n = ku_{n-1}$, k constant, $n \geq 2$,
then its solution is $u_n = k^{n-1}u_1$.

Example

Solve $u_n = 5u_{n-1}$, where $u_1 = 2$, and find u_5.

Solution

$$u_n = 5^{n-1}u_1$$

$$= 2 \times 5^{n-1}$$

$$\Rightarrow \quad u_5 = 2 \times 5^4$$

$$= 1250.$$

If the equation is of the type $u_n = ku_{n-1} + c$ then a general solution
can be found as follows :

$$u_n = ku_{n-1} + c$$

$$= k(ku_{n-2} + c) + c$$

$$= k^2 u_{n-2} + kc + c$$

$$= k^2(ku_{n-3} + c) + kc + c$$

$$\ldots \quad \ldots \quad \ldots$$

$$= k^{n-1}u_1 + k^{n-2}c + k^{n-3}c + \ldots + kc + c$$

$$= k^{n-1}u_1 + c(1 + k + k^2 + \ldots + k^{n-2}).$$

Again, there is a GP, $1 + k + k^2 + \ldots + k^{n-2}$, to be summed. This has a first term of 1, $n-1$ terms and a common ratio of k. So, provided that $k \neq 1$, the sum is

$$\frac{1\left(k^{n-1}-1\right)}{k-1}.$$

So if the difference equation is of the form $u_n = ku_{n-1} + c$, where k and c are constant, and $k \neq 1$, then it has solution

$$\boxed{u_n = k^{n-1}u_1 + \frac{c\left(k^{n-1}-1\right)}{k-1}}$$

The case $k = 1$ will be dealt with in a moment.

Example

Solve $u_n = 2u_{n-1} - 3$, $n \geq 2$, given $u_1 = 4$.

Solution

Using the formula above,

$$u_n = 2^{n-1} \times 4 - \frac{3\left(2^{n-1}-1\right)}{2-1}$$

$$= 4 \times 2^{n-1} - 3 \times 2^{n-1} + 3$$

$$= 2^{n-1} + 3.$$

You can see this formula works by finding, say, the value of u_2 from the difference equation as well.

Using the difference equation,

$$u_2 = 2u_1 - 3$$

$$= 2 \times 4 - 3$$

$$= 8 - 3$$

$$= 5.$$

Using the formula,

$$u_2 = 2^1 + 3$$

$$= 5.$$

Special case

In the formula for u_n, k cannot equal one. In this case, when $k = 1$, the difference equation is of the type

$$u_n = u_{n-1} + c, \quad n \geq 2.$$

This has the simple solution

$$\boxed{u_n = u_1 + (n-1)c}$$

This is the type of sequence in the parachute jump problem of Section 10.0. If, in that example, you let u_n be the number of metres fallen after n seconds, then

$$u_n = u_{n-1} + 10, \quad u_1 = 5$$

$$\Rightarrow \quad u_n = 10(n-1) + 5$$

For example,

$$u_4 = 10 \times 3 + 5$$

$$= 30 + 5$$

$$= 35.$$

You can now use the formula to solve the original problem of how long it takes to fall 600 metres.

Notation

In some cases, it is convenient to number the terms of a sequence,

$$u_0, u_1, u_2 \ldots$$

rather than

$$u_1, u_2, u_3, \ldots$$

This will often be the case in the next chapter. The different numbering affects both the difference equation and its solution.

For example, look again at the triangle numbers

$$1, 3, 6, 10, 15, \ldots$$

If these are denoted by u_1, u_2, u_3, ..., then, for example, $u_3 = 6$, $u_4 = 10$, $u_4 = u_3 + 4$, and in general $u_n = u_{n-1} + n$. The solution turns out to be

$$u_n = \tfrac{1}{2}n(n+1).$$

If, however, you start again and instead denote that sequence of triangle numbers by u_0, u_1, u_2, ..., then $u_2 = 6$, $u_3 = 10$, $u_3 = u_2 + 4$, in general $u_n = u_{n-1} + (n+1)$. In this case the solution is given by

$$u_n = \tfrac{1}{2}(n+1)(n+2)$$

(which, of course, could be obtained from the previous general solution by replacing n by $n+1$).

If you choose to write sequences beginning with the term u_0, then the solutions to difference equations of the form

$$u_n = ku_{n-1} + c$$

alter somewhat as shown below.

If $\qquad u_n = ku_{n-1} + c, \quad n \geq 1,$

then $\qquad \boxed{u_n = k^n u_0 + \dfrac{c(k^n - 1)}{k-1}, \quad k \neq 1}$

or, if $u_n = u_{n-1} + c$, then $u_n = u_0 + nc$.

Example

Find the solution of the equation $u_n = 3u_{n-1} + 4$, given $u_0 = 2$.

Solution

$$\begin{aligned} u_n &= 3^n u_0 + \frac{4(3^n - 1)}{3-1} \\ &= 3^n \times 2 + 2(3^n - 1) \\ &= 2 \times 3^n + 2 \times 3^n - 2 \\ &= 4 \times 3^n - 2. \end{aligned}$$

In the example above the **particular solution** to the difference equation $u_n = 3u_{n-1} + 4$ when $u_0 = 2$ has been found.

If you had not substituted for the value of u_0 (perhaps not knowing u_0) then a **general solution** could have been given as

$$u_n = 3^n u_0 + 2(3^n - 1).$$

This solution is valid for all sequences which have the same difference equation whatever the initial term u_0.

Example

Find the general solution of the difference equation

$$u_n = u_{n-1} + 4, \quad n \geq 1.$$

Solution

$$u_n = u_0 + 4n.$$

Exercise 10C

1. Write down the general solutions of :

    (a) $u_n = 4u_{n-1} + 2, \quad n \geq 2$

    (b) $u_n = 4u_{n-1} + 2, \quad n \geq 1$

    (c) $u_n = 3u_{n-1} - 5, \quad n \geq 1$

    (d) $u_{n+1} = u_n + 6, \quad n \geq 0$

    (e) $u_n = u_{n-1} - 8, \quad n \geq 2$

    (f) $u_n = -2u_{n-1} + 4, \quad n \geq 1$

    (g) $u_n + 3u_{n-1} - 2 = 0, \quad n \geq 1$

    (h) $u_n + 4u_{n-1} + 3 = 0, \quad n \geq 1$

    (i) $u_n = 4u_{n-1}, \quad n \geq 2$

    (j) $u_{n+1} = 4u_n - 5, \quad n \geq 0$

2. Find the particular solutions of these equations :

    (a) $u_0 = 1$ and $u_n = 3u_{n-1} + 5, \quad n \geq 1$

    (b) $u_1 = 3$ and $u_n = -2u_{n-1} + 6, \quad n \geq 2$

    (c) $u_0 = 4$ and $u_n = u_{n-1} - 3, \quad n \geq 1$

    (d) $u_1 = 0$ and $u_{n+1} = 5u_n + 3, \quad n \geq 1$

    (e) $u_0 = 3$ and $u_{n+1} = u_n + 7, \quad n \geq 0$

    (f) $u_0 = 1$ and $u_n + 3u_{n-1} = 1, \quad n \geq 1.$

10.4 Loans

Activity 4

Find out about the repayments on a loan from a bank, building society or other lending agency. You will need to know the rate of interest per annum, the term of the loan and the frequency of the repayment.

Use these variables in order to set up a difference equation :

u_n = amount in £ owing after n repayments

k = interest multiplier

c = repayment in £

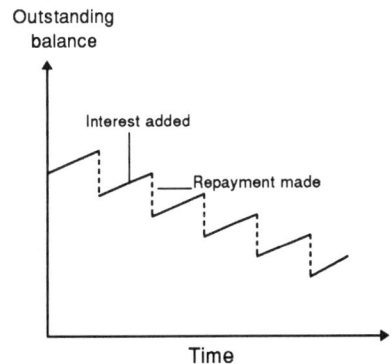

It will be of the type described in Section 10.2.

Solve your equation and then evaluate c, the repayment.
Remember that there is nothing owing after the final repayment
has been made.

How does your result compare with the repayments specified by
the lending agency?

The activity is quite difficult. So if you need help an example of a
similar type follows.

Example

Find the monthly repayment on a £500 loan at an interest rate of
24% p.a. over 18 months.

Solution

Let u_n be the amount owing after n months, so $u_0 = £500$ and c is
the repayment each month. You are given that the interest rate per
annum is 24%, so the interest rate per month is assumed to be 2%.
(Can you see why this is an approximation?) Then

$$u_n = u_{n-1} + (2\% \text{ of } u_{n-1}) - c.$$

Here, the interest has been added to the previous outstanding loan,
and one month's repayment subtracted.

$$\Rightarrow \quad u_n = u_{n-1} + 0.02 u_{n-1} - c$$

$$u_n = 1.02 u_{n-1} - c.$$

Remembering that you started with u_0, and noting the switch to a
negative c, the general solution of this type of equation is

$$u_n = k^n u_0 + \frac{c(k^n - 1)}{(k-1)}$$

where in this case $k = 1.02$.

So
$$u_n = (1.02)^n u_0 - \frac{c(1.02^n - 1)}{1.02 - 1}$$

$$= 1.02^n u_0 - 50c(1.02^n - 1)$$

or
$$u_n = 1.02^n u_0 - 50c(1.02^n - 1).$$

At the end of the term $n = 18$, so

$$u_{18} = 1.02^{18} u_0 - 50c\left(1.02^{18} - 1\right).$$

But $u_0 = 500$ and $u_{18} = 0$ (the loan has been paid off)

$$\Rightarrow \quad 0 = 500 \times 1.02^{18} - 50c\left(1.02^{18} - 1\right)$$

$$\Rightarrow \quad c = \frac{500 \times 1.02^{18}}{50\left(1.02^{18} - 1\right)}$$

$$\approx \frac{714.123}{21.4123}$$

$$\approx £33.35.$$

Repayments are made at the rate of £33.35 per month.

In the example on the previous page, as in many situations involving loans, the monthly rate is taken as one twelfth of the annual rate. Is the rate of 24% p.a. justifiable as a measure of the interest paid? What is the significance of an APR?

Exercise 10D

1. (a) Find the general solution of
    $$u_n = 3u_{n-1} + 4, n \geq 2$$

    (b) Find the general solution of
    $$u_n = \tfrac{1}{2}u_{n-1} + 2, \; n \geq 2$$

    (c) Solve $q_n = q_{n-1} + 3$, given $q_3 = 7$, $n \geq 2$

    (d) Solve $a_n - 2a_{n-1} = 0$, given $a_1 = 4$, $n \geq 2$

    (e) Solve $b_n = 4b_{n-1} + 5$, given $b_1 = 2$, $n \geq 2$

2. Form and solve the difference equation associated with the sequence
    7, 17, 37, 77, 157 ...

3. Find the monthly repayment on a £400 loan over a period of $1\tfrac{1}{2}$ years at an interest rate of 15% p.a.

4. Find the monthly repayment on a £2000 loan over a term of 3 years at an interest rate of 21% p.a.

5. Write a computer program to solve the type of problem in Question 3.

6. A steel works is increasing production by 1% per month from a rate of 2000 tonnes per month. Orders (usage of the steel) remain at 1600 tonnes per month. How much steel will be stock-piled after periods of 12 months and 2 years?

7. A loan of £1000 is taken out at an interest rate of 24% p.a. How long would it take to repay the loan at a rate of £50 per month.

    [**Note**: this problem is similar to Question 3, except that the value of n cannot easily be found after you have solved the difference equation. A search technique on your calculator or a simple computer program will be necessary.]

10.5 Non-homogenous linear equations

So far you have met equations of the form

$$u_n = ku_{n-1}, \quad k \text{ constant}$$

This is called a **homogeneous** equation, involving terms in u_n, and u_{n-1} only.

You have also solved equations of the form

$$u_n = ku_{n-1} + c, \quad k, c \text{ constant}$$

This is a **non-homogeneous** equation, due to the extra term c, but k and c are still constant.

Earlier you met the difference equation associated with the triangle numbers

$$u_n = u_{n-1} + n$$

This is again a non-homogeneous equation, but the term n is not constant.

In general, equations of the type $u_n = ku_{n-1} + f(n)$ are difficult to solve. Expanding u_n as a series is occasionally successful - it depends usually on how complicated $f(n)$ is. In the next chapter you will meet two methods for solving these equations, and also second order equations of a similar type (these involve terms in u_n, u_{n-1} and u_{n-2}). One method is based on trial and error and the other on the use of **generating functions**. Simpler equations such as the one for the triangle numbers can be solved quite quickly by expanding u_n as a series. Thus

$$u_n = (u_{n-2} + n - 1) + n$$

$$= u_{n-2} + (n-1) + n$$

$$= (u_{n-3} + n - 2) + (n-1) + n$$

$$= \dots \quad \dots \quad \dots \quad \dots \quad \dots$$

$$= 1 + 2 + 3 + \dots + (n-1) + n.$$

This is an **arithmetic progression** (AP). Summing gives

$$u_n = \frac{n}{2}(n+1).$$

You can check this result by evaluating, say, u_5.

Activity 5

Try out this method on one of the 'dot' patterns which you designed in Activity 2.

The problems in the following exercise should all yield to a method similar to the one above. The function $f(n)$ will be restricted to the form n, n^2 or k^n, where k is a constant. This will also be true of later difference equations in the next chapter.

Exercise 10E

1. Solve $u_n = u_{n-1} + n$ if $u_1 = 5$.

2. Find the general solution in terms of u_1 :

    (a) $u_n = u_{n-1} + n^2$ (b) $u_n = u_{n-1} + 2^n$

    *(c) $u_n = 2u_{n-1} + n$

3. If $u_n = ku_{n-1} + 5$ and $u_1 = 4$, $u_2 = 17$ find the values of k and u_6.

4. The productivity of an orchard of 2000 trees increases by 5% each year due to improved farming techniques. The farmer also plants a further 100 trees per year. Estimate the percentage improvement in productivity during the next 10 years.

5. A person saves by putting £50 each month from their salary in the bank at an interest rate of 12% p.a. How much interest should accrue during one year?

10.6 A population problem

Changes in population can often be modelled using difference equations. The underlying problems are similar to some of the financial problems you have met in this section.

What factors influence demographic change?

Activity 6

The birth and death rates in a country are 40 per thousand and 15 per thousand per year respectively. The initial population is 50 million.

Form a difference equation which gives the population at the end of a year in relation to that at the end of the previous year. Solve the equation and estimate the population in 10 years time.

If, due to the high birth rate, emigration takes place at a rate of 10000 per year, how will this change your results?

Population analysis can be refined by taking into account many more factors than in the above problem. In particular, the population pyramid shows that different age group sizes affect the whole population in different ways.

Suppose the population of a country is split into two age groups:

Group 1   consisting of the 0 - 12 year olds, and

Group 2   consisting of the rest,

and assume that births only occur in Group 2. Each group will have its own death rate.

Define $p_1(t)$ as the population of the 0 - 12 group in year t

$p_2(t)$ as the population of the 13+ group in year t

b as the birth rate

d_1 as the death rate in the 0 - 12 group

d_2 as the death rate in the 13+ group.

One further assumption made is that in each year one twelfth of the survivors from Group 1 progress to Group 2.

For **Group 1**

$$p_1(t+1) \;=\; \underbrace{b\,p_2(t)}_{\substack{\text{those born} \\ \text{to people in} \\ \text{Group 2}}} \;+\; \underbrace{\tfrac{11}{12}p_1(t)\left(1-d_1\right)}_{\substack{\text{of those who survive,} \\ \tfrac{11}{12}\text{ remain in Group 1}}}$$

Remember that $\frac{1}{12}$ of the survivors of Group 1 transfer to Group 2 each year.

For **Group 2**

$$p_2(t+1) = \tfrac{1}{12}p_1(t)(1-d_1) + p_2(t)(1-d_2)$$

Using matrices, both equations can be more simply written as

$$\begin{bmatrix} p_1(t+1) \\ p_2(t+1) \end{bmatrix} = \begin{bmatrix} \tfrac{11}{12}(1-d_1) & b \\ \tfrac{1}{12}(1-d_1) & 1-d_2 \end{bmatrix} \begin{bmatrix} p_1(t) \\ p_2(t) \end{bmatrix}$$

Let $P_t \equiv \begin{bmatrix} p_1(t) \\ p_2(t) \end{bmatrix}$

and
$$A = \begin{bmatrix} \frac{11}{12}(1-d_1) & b \\ \frac{1}{12}(1-d_1) & 1-d_2 \end{bmatrix}$$

then
$$P_{t+1} = AP_t.$$

This is a difference equation using matrices! Using the earlier solution to this type of equation you can see that

$$P_t = A^t P_0$$

where P_0 is the initial population .

Suppose that $p_1(0) = 5$ million, $p_2(0) = 15$ million , and that the population parameters are

$$b = 0.4$$

$$d_1 = 0.016$$

$$d_2 = 0.03$$

then the above solution for P_t becomes

$$P_t = \begin{bmatrix} \frac{11}{12}(1-0.016) & 0.04 \\ \frac{1}{12}(1-0.016) & 1-0.03 \end{bmatrix}^t \begin{bmatrix} 5 \\ 15 \end{bmatrix}$$

$$= \begin{bmatrix} 0.902 & 0.04 \\ 0.082 & 0.97 \end{bmatrix}^t \begin{bmatrix} 5 \\ 15 \end{bmatrix}$$

For example, the population after one year can be calculated by simple matrix multiplication :

$$P_1 = \begin{bmatrix} 0.902 & 0.04 \\ 0.082 & 0.97 \end{bmatrix} \begin{bmatrix} 5 \\ 15 \end{bmatrix}$$

$$= \begin{bmatrix} 5.11 \\ 14.96 \end{bmatrix}$$

The total population is therefore 20.07 million.

If the population in 10 years is required, then it is fairly straightforward to evaluate A^{10} using a simple program for matrix multiplication with a computer or a modern graphic/programmable calculator.

Activity 7

Produce a simple program which will multiply matrices as required in the above example. Use it to evaluate the population in 10 years time.

Activity 8

Find a population pyramid with associated birth and death rates for a country of your choice and model the population growth (or decay) as in the above example.

You may wish to consider more than two age groupings. This may prove quite difficult and you will need to adapt your program to multiply larger matrices.

10.7 Miscellaneous Exercises

1. Find the general solution to these difference equations in terms of u_1:

   (a) $u_n = 2u_{n-1}$ (b) $u_n - 3u_{n-1} = 3$

   (c) $u_n - 3u_{n-1} = n$

2. By writing down a difference equation and solving it, find the tenth term of this sequence:

   2 4 10 28 82 244.

3. Find the monthly repayment on a loan of £600 over a period of 12 months at a rate of interest of 3% per month.

4. The population of a country is $12\frac{1}{2}$ million. The birth rate is 0.04, the death rate is 0.03 and 50 000 immigrants arrive in the country each year. Estimate the population in 20 years' time.

5. In a round robin tournament every person (or team) plays each of the others. If there are n players, how many more games are needed if one more player is included. Use this result to set up a difference equation, solve it, and then evaluate the number of games needed for 20 players (this confirms the answer found to Question 3(b) in Exercise 6A).

6. If $u_n = pu_{n-1} + q$, $n \geq 1$, and $u_1 = 2$, $u_2 = 3$, $u_3 = 7$, find the value of u_6.

7. Compare the monthly repayment on a mortgage of £30 000 at an interest rate of 12% p.a. over 25 years for the two standard types of mortgage:

   (a) a **repayment** mortgage, where you pay a fixed amount each month, the interest is calculated each month on the remaining debt, and the amount of repayment is calculated so that the debt is paid off after 25 years;

   (b) an **endowment** mortgage where the loan stays at £30 000 for the whole of the 25 years, you pay the interest on that and an additional £40 a month for a sort of insurance policy known as an endowment policy. At the end of the 25 years the insurance policy matures and the insurance firm pays off the debt.

   (You will find that the latter costs more, but in practice when the insurance policy matures it pays well over the £30 000, thus giving you an additional lump sum back.)

8. A population of 100 million can be divided into age groups. **Group 1**, 0-16 years, has a death rate of 0.025 (no birth rate) and a population of 60 million. **Group 2**, 17+ years, has a birth rate of 0.04 and a death rate of 0.03. Investigate the growth/decay of the population. Make a prediction for the population size in 3 years' time.

9. A person is repaying a loan of £5000 at £200 per month. The interest rate is 3% per month. How long will it take to repay the loan?

10. Within a population of wild animals the birth rate is 0.2, while the death rate is 0.4. Zoos worldwide are attempting to reintroduce animals. At present the population is 5000 and 100 animals per year are being introduced. The rate of increase of introduction is 20% per year. Will the animals survive in the long run?

11. Investigate the problem of the Tower of Hanoi with four needles for the rings.

11 DIFFERENCE EQUATIONS 2

Objectives

After studying this chapter you should

* be able to obtain the solution of any linear homogeneous second order difference equation;

* be able to apply the method of solution to contextual problems;

* be able to use generating functions to solve non-homogeneous equations.

11.0 Introduction

In order to tackle this chapter you should have studied a substantial part of the previous chapter on first order difference equations. The problems here deal with rather more sophisticated equations, called second order difference equations, which derive from a number of familiar contexts. This is where the rabbits come in.

It is well known that rabbits breed fast. Suppose that you start with one new-born pair of rabbits and every month any pair of rabbits gives birth to a new pair, which itself becomes productive after a period of two months. How many rabbits will there be after n months? The table shows the results for the first few months.

Month	1	2	3	4	5	6
No. of pairs (u_n)	1	1	2	3	5	8

The sequence u_n is a famous one attributed to a 13th century mathematician *Leonardo Fibonacci*. As you can see the next term can be found by adding together the previous two.

The nth term u_n can be written as

$$u_n = u_{n-1} + u_{n-2}$$

and difference equations like this with terms in u_n, u_{n-1} and u_{n-2} are said to be of the **second order** (since the difference between n and $n-2$ is 2).

Activity 1   Fibonacci numbers

The **Fibonacci numbers** have some remarkable properties. If you divide successive terms by the previous term you obtain the sequence,

$$\frac{1}{1}, \frac{2}{1}, \frac{3}{2}, \frac{5}{3}, \ldots = 1, \ 2, \ 1.5, \ 1.\dot{6}, \ldots$$

Continue this sequence, say to the 20th term, and find its reciprocal. What do you notice? Can you find an equation with a solution which gives you the limit of this sequence?

Exercise 11A

1. If $u_n = 2u_{n-1} + u_{n-2}$ and $u_1 = 2$, $u_2 = 5$, find the values of u_3, u_4, u_5.

2. $u_n = pu_{n-1} + qu_{n-2}$ describes the sequence 1, 2, 8, 20, 68, ... Find p and q.

3. If F_n is a term of the Fibonacci sequence, investigate the value of $F_{n+1} F_{n-1} - F_n{}^2$.

4. What sequences correspond to the difference equation $u_n = u_{n-1} - u_{n-2}$, $n \geq 3$? Choose your own values for u_1 and u_2.

5. Find the ratio of the length of a diagonal to a side of a regular pentagon. What do you notice?

6. Investigate the limit of $\dfrac{u_n}{u_{n-1}}$ if $u_n = u_{n-1} + 2u_{n-2}$.

7. Show that $F_1 + F_2 + F_3 + \ldots + F_n = F_{n+2} - 1$, where F_n is a term of the Fibonacci sequence.

11.1   General solutions

When you solved difference equations in the previous chapter, any **general** solution had an unknown constant left in the solution. Usually this was u_1 or u_0.

Example

Solve $u_n = 4u_{n-1} - 3$.

Solution

$$
\begin{aligned}
u_n &= 4^n u_0 - \frac{3\left(4^n - 1\right)}{3} \\[2mm]
&= 4^n u_0 - \left(4^n - 1\right) \\[2mm]
&= 4^n \left(u_0 - 1\right) + 1
\end{aligned}
$$

Alternatively you could write

$$u_n = A4^n + 1, \text{ replacing } u_0 - 1 \text{ by A.}$$

This first order equation has one arbitrary constant in its general solution. Knowing the value of u_0 would give you a **particular solution** to the equation.

Say $u_0 = 4$ then

$$4 = A.4^0 + 1 = A + 1.$$

So $A = 3$ and

$$u_n = 3 \times 4^n + 1.$$

In a similar way, **general solutions** to second order equations have two arbitrary constants. Unfortunately, an iterative technique does not work well for these equations, but, as you will see, a guess at the solution being of a similar type to that for first order equations does work.

Suppose

$$\boxed{u_n = pu_{n-1} + qu_{n-2}} \tag{1}$$

where p, q are constants, $n \geq 2$.

This is a **second order homogeneous linear difference equation** with constant coefficients.

As a solution, try $u_n = Am^n$, where m and A are constants. This choice has been made because $u_n = k^n u_0$ was the solution to the first order equation $u_n = ku_{n-1}$.

Substituting $u_n = Am^n$, $u_{n-1} = Am^{n-1}$ and $u_{n-2} = Am^{n-2}$ into equation (1) gives

$$Am^n = Apm^{n-1} + Aqm^{n-2}$$

$$\Rightarrow \quad Am^{n-2}\left(m^2 - pm - q\right) = 0.$$

If $m = 0$, or $A = 0$, then equation (1) has trivial solutions (i.e. $u_n = 0$). Otherwise, if $m \neq 0$ and $A \neq 0$ then

$$m^2 - pm - q = 0.$$

This is called the **auxiliary equation** of equation (1).

It has the solution

$$m_1 = \frac{p + \sqrt{p^2 + 4q}}{2} \quad \text{or} \quad m_2 = \frac{p - \sqrt{p^2 + 4q}}{2}.$$

m_1 and m_2 can be real or complex. The case where $m_1 = m_2$ is special, as you will see later.

Suppose for now that $m_1 \neq m_2$, then it has been shown that both $u_n = Am_1^n$ and $u_n = Bm_2^n$ are solutions of (1), where A and B are constants.

Can you suggest the form of the general solution?

In fact it is easy to show that a **linear** combination of the two solutions is also a solution. This follows since both Am_1^n and Bm_2^n satisfy equation (1) giving

$$Am_1^n = Am_1^{n-1}p + Am_1^{n-2}q$$

and $\qquad Bm_2^n = Bm_2^{n-1}p + Bm_2^{n-2}q$

$$\Rightarrow \quad Am_1^n + Bm_2^n = p\left(Am_1^{n-1} + Bm_2^{n-1}\right) + q\left(Am_1^{n-2} + Bm_2^{n-2}\right).$$

So $Am_1^n + Bm_2^n$ is also a solution of equation (1), and can in fact be shown to be the general solution. That is **any** solution of (1) will be of this form.

In summary, the general solution of $u_n = pu_{n-1} + qu_{n-2}$ is

$$\boxed{u_n = Am_1^n + Bm_2^n, \quad m_1 \neq m_2}$$

where A, B are arbitrary constants and m_1, m_2 are the solutions of the auxiliary equation $m^2 - pm - q = 0$.

Example

Find the general solution of $u_n = 2u_{n-1} + 8u_{n-2}$.

Solution

The auxiliary equation is $m^2 - 2m - 8 = 0$.

This has solutions $m_1 = 4$ and $m_2 = -2$.

The general solution is therefore

$$u_n = A4^n + B(-2)^n.$$

Example

Solve $u_n + 3u_{n-2} = 0$, $n \geq 3$, given that $u_1 = 1$ and $u_2 = 3$.

Solution

The auxiliary equation is $m^2 + 3 = 0$.

$$\Rightarrow \quad m^2 = -3$$

$$\Rightarrow \quad m_1 = \sqrt{3}\,i \quad \text{and} \quad m_2 = -\sqrt{3}\,i \quad (\text{where } i = \sqrt{-1}).$$

The **general solution** to the equation is therefore

$$u_n = A\left(\sqrt{3}\,i\right)^n + B\left(-\sqrt{3}\,i\right)^n.$$

When $n = 1$, $u_1 = 1$ and since $u_1 = A\left(\sqrt{3}\,i\right)^1 + B\left(-\sqrt{3}\,i\right)^1$

$$\Rightarrow \quad 1 = A\sqrt{3}\,i - B\sqrt{3}\,i$$

$$\frac{1}{\sqrt{3}\,i} = A - B. \qquad (2)$$

When $n = 2$, $u_2 = 3$ and $u_2 = A\left(\sqrt{3}\,i\right)^2 + B\left(-\sqrt{3}\,i\right)^2$

$$\Rightarrow \quad 3 = -A3 - B3$$

$$\Rightarrow \quad -1 = A + B. \qquad (3)$$

Adding (2) and (3) gives

$$2A = -1 + \frac{1}{\sqrt{3}\,i} = -1 - \frac{i}{\sqrt{3}}$$

$$\Rightarrow \quad A = -\frac{1}{2}\left(1 + \frac{i}{\sqrt{3}}\right)$$

and

$$B = -\frac{1}{2}\left(1 - \frac{i}{\sqrt{3}}\right).$$

Thus the particular solution to the equation, for $u_1 = 1$, $u_2 = 3$, is given by

$$u_n = -\frac{1}{2}\left(1 + \frac{i}{\sqrt{3}}\right)\left(\sqrt{3}\,i\right)^n - \frac{1}{2}\left(1 - \frac{i}{\sqrt{3}}\right)\left(-\sqrt{3}\,i\right)^n \qquad (4)$$

Although this solution is given in terms of the complex number $i = \sqrt{-1}$, it is in fact always a real number.

Activity 2

Show that equation (4) gives $u_1 = 1$ and $u_2 = 3$. Also use this equation to evaluate u_3 and u_4, and check these answers directly from the original difference equation, $u_n + 3u_{n-2} = 0$.

Exercise 11B

1. Find the general solutions to

   (a) $u_n = u_{n-1} + 6u_{n-2}$

   (b) $u_n = 4u_{n-1} + u_{n-2}$

   (c) $u_n - u_{n-1} - 2u_{n-2} = 0$.

2. Find the general solution of the difference equation associated with the Fibonacci sequence. Use $u_0 = 1$, $u_1 = 1$, to find the particular solution.

3. Solve $u_n + 4u_{n-2} = 0$, $n \geq 3$, if $u_1 = 2$, $u_2 = -4$.

4. Solve $u_n - 6u_{n-1} + 8u_{n-2} = 0$, $n \geq 3$, given $u_1 = 10$, $u_2 = 28$. Evaluate u_6.

5. Find the nth term of the sequence

   $-3, 21, 3, 129, 147 \ldots$

11.2 Equations with equal roots

When $m_1 = m_2$, the solution in Section 11.1 would imply that

$$u_n = Am_1^n + Bm_1^n$$

$$= m_1^n(A+B)$$

$$= m_1^n C, \text{ where } C = A + B.$$

In this case there is really only one constant, compared with the two expected. Trials show that another possibility for a solution to $u_n = pu_{n-1} + qu_{n-2}$ is $u_n = Dnm_1^n$, and as you will see below, this solution, combined with one of the form Cm_1^n, gives a general solution to the equation when $m_1 = m_2$.

If $u_n = Dnm_1^n$ then

$$u_{n-1} = D(n-1)m_1^{n-1}$$

and $\quad u_{n-2} = D(n-2)m_1^{n-2}$.

If $u_n = Dnm_1^n$ is a solution of (1), then $u_n - pu_{n-1} - qu_{n-2}$ should equal zero.

$$u_n - pu_{n-1} - qu_{n-2}$$

$$= Dnm_1^n - pD(n-1)m_1^{n-1} - qD(n-2)m_1^{n-2}$$

$$= Dm_1^{n-2}\left[nm_1^2 - (n-1)pm_1 - (n-2)q\right]$$

$$= Dm_1^{n-2}\left[n\left(m_1^2 - pm_1 - q\right) + pm_1 + 2q\right]$$

$$= Dm_1^{n-2}\left(pm_1 + 2q\right)$$

because $m_1^2 - pm_1 - q = 0$.

Now, the auxiliary equation has equal roots, which means that

$$p^2 + 4q = 0 \quad \text{and} \quad m_1 = \frac{p}{2}.$$

Therefore $\quad u_n - pu_{n-1} - qu_{n-2} = Dm_1^{n-2}\left(p \times \frac{p}{2} + 2q\right)$

$$= 2Dm_1^{n-2}\left(p^2 + 4q\right)$$

$$= 0, \text{ since } p^2 + 4q = 0.$$

So $u_n = Dnm_1^n$ is a solution and therefore $Dnm_1^n + Cm_1^n$ will be also. This can be shown by using the same technique as for the case when $m_1 \neq m_2$.

In summary, when $p^2 + 4q = 0$ the general solution of $u_n = pu_{n-1} + qu_{n-2}$ is

$$\boxed{u_n = Cm_1^n + Dnm_1^n}$$

where C and D are arbitrary constants.

Example

Solve $u_n + 4u_{n-1} + 4u_{n-2} = 0$, $n \geq 3$, if $u_1 = -2$ and $u_2 = 12$.
Evaluate u_5.

Solution

The auxiliary equation is

$$m^2 + 4m + 4 = 0$$

$$\Rightarrow \quad (m+2)^2 = 0$$

$$\Rightarrow \quad m_1 = m_2 = -2.$$

Therefore the general solution is

$$u_n = Dn(-2)^n + C(-2)^n$$

or $\quad\quad u_n = (-2)^n(C + Dn).$

If $\quad\quad u_1 = -2, \quad -2(C+D) = -2 \quad \Rightarrow \quad C + D = 1.$

Also, as $\quad u_2 = 12, \quad 4(C+2D) = 12 \quad \Rightarrow \quad C + 2D = 3.$

These simultaneous equations can be solved to give $C = -1$ and $D = 2$.

Thus $\quad\quad u_n = (-2)^n(2n-1)$

and $\quad\quad u_5 = (-2)^5(10-1)$

$$= -32 \times 9$$

$$= -288.$$

Activity 3

Suppose that a pair of mice can produce two pairs of offspring every month and that mice can reproduce two months after birth. A breeder begins with a pair of new-born mice. Investigate the number of mice he can expect to have in successive months. You will have to assume no mice die and pairs are always one female and one male!

If a breeder begins with ten pairs of mice, how many can he expect to have bred in a year?

Exercise 11C

1. Find the general solutions of
   (a) $u_n - 4u_{n-1} + 4u_{n-2} = 0$
   (b) $u_n = 2u_{n-1} - u_{n-2}$.

2. Find the particular solution of
   $u_n - 6u_{n-1} + 9u_{n-2} = 0$, $n \geq 3$, when $u_1 = 9$, $u_2 = 36$.

3. If $u_1 = 0, u_2 = -4$, solve $u_{n+2} + u_n = 0$, $n \geq 1$, giving u_n in terms of i.

4. Find the particular solution of
   $u_{n+2} + 2u_{n+1} + u_n = 0$, $n \geq 1$, when $u_1 = -1$, $u_2 = -2$.

5. Find the solutions of these difference equations.
   (a) $u_n - 2u_{n-1} - 15u_{n-2} = 0$, $n \geq 3$, given $u_1 = 1$ and $u_2 = 77$.
   (b) $u_n = 3u_{n-2}$, $n \geq 3$, given $u_1 = 0$, $u_2 = 3$.
   (c) $u_n - 6u_{n-1} + 9u_{n-2} = 0$, $n \geq 3$, given $u_1 = 9$, $u_2 = 45$.

6. Form and solve the difference equation defined by the sequence in which the nth term is formed by adding the previous two terms and then doubling the result, and in which the first two terms are both one.

11.3 A model of the economy

In good times, increased national income will promote increased spending and investment.

If you assume that government expenditure is constant (G) then the remaining spending can be assumed to be composed of investment (I) and private spending on consumables (P). So you can model the national income (N) by the equation

$$N_t = I_t + P_t + G, \text{ where } t \text{ is the year number} \qquad (1)$$

If income increases from year $t-1$ to year t, then you would assume that private spending will increase in year t proportionately. So you can write :

$$P_t = AN_{t-1}, \text{ where } A \text{ is a constant.}$$

Also, extra private spending should promote additional investment. So you can write :

$$I_t = B(P_t - P_{t+1}), \text{ where } B \text{ is a constant.}$$

Substituting for P_t and I_t in (1) gives

$$N_t = AN_{t-1} + B(P_t - P_{t-1}) + G$$

$$= AN_{t-1} + B(AN_{t-1} - AN_{t-2}) + G$$

$$N_t = A(B+1)N_{t-1} - ABN_{t-2} + G. \qquad (2)$$

So far you have not met equations of this type in this chapter. It is a second order difference equation, but it has an extra constant G.

Before you try the activity below, discuss the effects you think the values of A and B will have on the value of N as t increases.

Activity 4

Take, as an example, an economy in which for year 1, $N_1 = 2$ and for year 2, $N_2 = 4$. Suppose that $G = 1$. By using the difference equation (2) above, investigate the change in the size of N over a number of years for different values of A and B.

At this stage you should not attempt an algebraic solution!

Equation (2) is an example of a **non-homogeneous difference equation. Homogeneous** second order equations have the form

$$\boxed{u_n + au_{n-1} + bu_{n-2} = 0}$$

There are no other terms, unlike equation (2) which has an additional constant G.

11.4 Non-homogeneous equations

You have seen how to solve homogeneous second order difference equations; i.e. ones of the form given below but where the right-hand side is zero. Turning to non-homogeneous equations of the form

$$\boxed{u_n + au_{n-1} + bu_{n-2} = f(n)}$$

where f is a function of n, consider as a first example the equation

$$6u_n - 5u_{n-1} + u_{n-2} = n, \quad (n \geq 3) \tag{1}$$

Activity 5

Use a computer or calculator to investigate the sequence u_n defined by

$$6u_n - 5u_{n-1} + u_{n-2} = n, \quad (n \geq 3)$$

for different starting values. Start, for example, with $u_1 = 1$, $u_2 = 2$ and then vary either or both of u_1 and u_2. How does the sequence behave when n is large?

From the previous activity, you may have had a feel for the behaviour or structure of the solution. Although its proof is beyond the scope of this text, the result can be expressed as

$$u_n = \begin{pmatrix} \text{general solution of} \\ \text{associated homogeneous} \\ \text{equation} \end{pmatrix} + \begin{pmatrix} \text{one particular} \\ \text{solution of the} \\ \text{full equation} \end{pmatrix}$$

That is, to solve

$$6u_n - 5u_{n-1} + u_{n-2} = n, \quad (n \geq 2)$$

you first find the general solution of the associated homogeneous equation

$$6u_n - 5u_{n-1} + u_{n-2} = 0 \qquad (2)$$

and, to this, add one particular solution of the full equation.

You have already seen in Section 11.2 how to solve equation (2). The auxiliary equation is

$$6m^2 - 5m + 1 = 0$$

$$\Rightarrow \quad (3m-1)(2m-1) = 0$$

$$\Rightarrow \quad m = \tfrac{1}{3} \text{ or } \tfrac{1}{2}.$$

So the general solution of (2) is given by

$$u_n = A\left(\tfrac{1}{3}\right)^n + B\left(\tfrac{1}{2}\right)^n, \qquad (3)$$

where A and B are constants.

The next stage is to find one particular solution of the full equation (1).

Can you think what type of solution will satisfy the full equation?

In fact, once you have gained experience in solving equations of this type, you will recognise that u_n will be of the form

$$u_n = a + bn$$

(which is a generalisation of the function on the right-hand side, namely n).

So if $\quad u_n = a + bn$

$$\Rightarrow \quad u_{n-1} = a + b(n-1)$$

$$\Rightarrow \quad u_{n-2} = a + b(n-2)$$

and to satisfy (1), we need

$$6(a+bn) - 5(a+b(n-1)) + a + b(n-2) = n$$

$$6a + 6bn - 5a - 5bn + 5b + a + bn - 2b = n$$

$$2a + 3b + n(2b) = n.$$

Each side of this equation is a polynomial of degree 1 in n.

How can both sides be equal?

To ensure that it is satisfied for all values of n, equate cooefficients on each side of the equation.

$$\text{constant term} \quad \Rightarrow \quad 2a + 3b = 0$$

$$n \text{ term} \quad \Rightarrow \quad 2b = 1.$$

So $b = \frac{1}{2}$ and $a = -\frac{3}{4}$, and you have shown that one particular solution is given by

$$u_n = -\frac{3}{4} + \frac{1}{2}n. \tag{4}$$

To complete the general solution, add (4) to (3) to give

$$u_n = A\left(\frac{1}{3}\right)^n + B\left(\frac{1}{2}\right)^n - \frac{3}{4} + \frac{1}{2}n.$$

Activity 6

Find the solution to equation (1) which satisfies $u_1 = 1$, $u_2 = 2$.

The main difficulty of this method is that you have to 'guess' the form of the particular solution. The table below gives the usual form of the solution for various functions $f(n)$.

$f(n)$	Form of particular solutions
constant	a
n	$a + bn$
n^2	$a + bn + cn^2$
k^n	ak^n (or ank^n in special cases)

The next three examples illustrate the use of this table.

Example

Find the general solution of $6u_n - 5u_{n-1} + u_n = 2$.

Solution

From earlier work, the form of the general solution is

$$u_n = A\left(\frac{1}{3}\right)^n + B\left(\frac{1}{2}\right)^n + \left(\begin{array}{c}\text{one particular}\\\text{solution}\end{array}\right)$$

For the particular solution, try

$$u_n = a \quad \Rightarrow \quad u_{n-1} = a \quad \text{and} \quad u_{n-2} = a$$

which on subtituting in the equation gives

$$6a - 5a + a = 2 \implies a = 1.$$

Hence $u_n = 1$ is a particular solution and the general solution is given by

$$u_n = A\left(\tfrac{1}{3}\right)^n + B\left(\tfrac{1}{2}\right)^n + 1.$$

Example

Find the general solution of $6u_n - 5u_{n-1} + u_{n-2} = 2^n$.

Solution

For the particular solution try $u_n = a2^n$, so that $u_{n-1} = a2^{n-1}$, and substituting in the equation gives

$$6a2^n - 5a2^{n-1} + a2^{n-2} = 2^n$$

$$2^{n-2}(6a \times 4 - 5a \times 2 + a) = 2^n$$

$$24a - 10a + a = 4$$

$$\implies \quad a = \tfrac{4}{15}$$

and so the general solution is given by

$$u_n = A\left(\tfrac{1}{3}\right)^n + B\left(\tfrac{1}{2}\right)^n + \left(\tfrac{4}{15}\right)2^n.$$

In the next example, the equation is

$$6u_n - 5u_{n-1} + u_{n-2} = \left(\tfrac{1}{2}\right)^n.$$

Can you see why the usual trial for a particular solution, namely $u_n = a\left(\tfrac{1}{2}\right)^n$ will not work?

Example

Find the general solution of $6u_n - 5u_{n-1} + u_{n-2} = \left(\tfrac{1}{2}\right)^n$.

Solution

If you try $u_n = a\left(\tfrac{1}{2}\right)^n$ for a particular solution, you will not be able to find a value for the constant a to give a solution. This is because the term $B\left(\tfrac{1}{2}\right)^n$ is already in the solution of the associated homogeneous equation. In this special case, try

$$u_n = an\left(\tfrac{1}{2}\right)^n$$

so that $\qquad u_{n-1} = a(n-1)\left(\tfrac{1}{2}\right)^{n-1}$

and $\qquad u_{n-2} = a(n-2)\left(\tfrac{1}{2}\right)^{n-2}.$

Substituting in the equation gives

$$6an\left(\tfrac{1}{2}\right)^n - 5a(n-1)\left(\tfrac{1}{2}\right)^{n-1} + a(n-2)\left(\tfrac{1}{2}\right)^{n-2} = \left(\tfrac{1}{2}\right)^n$$

$$\left(\tfrac{1}{2}\right)^{n-2}\left(6an\left(\tfrac{1}{2}\right)^2 - 5a(n-1)\tfrac{1}{2} + a(n-2)\right) = \left(\tfrac{1}{2}\right)^n$$

$$\tfrac{3}{2}an - \tfrac{5}{2}an + \tfrac{5}{2}a + an - 2a = \tfrac{1}{4}$$

$$\tfrac{1}{2}a = \tfrac{1}{4} \quad \text{(the } n \text{ terms cancel out)}.$$

Hence $a = \tfrac{1}{2}$, and the particular solution is

$$u_n = \tfrac{1}{2}n\left(\tfrac{1}{2}\right)^n = n\left(\tfrac{1}{2}\right)^{n+1}.$$

The general solution is given by

$$u_n = A\left(\tfrac{1}{3}\right)^n + B\left(\tfrac{1}{2}\right)^n + n\left(\tfrac{1}{2}\right)^{n+1}.$$

The examples above illustrate that, although the algebra can become quite complex, the real problem lies in the intelligent choice of the form of the solution. Note that if the usual form does not work, then the degree of the polynomial being tried should be increased by one. The next two sections will illustrate a more general method, not dependent on inspired guesswork!

Exercise 11 D

1. Find the general solution of the difference equation $u_n - 5u_{n-1} + 6u_{n-2} = f(n)$ when

    (a) $f(n) = 2$

    (b) $f(n) = n$

    (c) $f(n) = 1 + n^2$

    (d) $f(n) = 5^n$

    (e) $f(n) = 2^n.$

2. Find the complete solution of

    $$u_n - 7u_{n-1} + 12u_{n-2} = 2^n$$

    when $u_1 = 1$ and $u_2 = 1$.

3. Find the general solution of

    $$u_n + 3u_{n-1} - 10u_{n-2} = 2^n$$

    and determine the solution which satisfies $u_1 = 2$, $u_2 = 1$.

11.5 Generating functions

This section will introduce a new way of solving difference equations by first applying the method to homogeneous equations.

A different way of looking at a sequence u_0, u_1, u_2, u_3 is as the coefficients of a power series

$$G(x) = u_0 + u_1 x + u_2 x^2 + \ldots$$

Notice that the sequence and series begin with u_0 rather than u_1. This makes the power of x and the suffix of u the same, and will help in the long run.

$G(x)$ is called the **generating function** for the sequence u_0, u_1, u_2, ... This function can be utilised to solve difference equations. Here is an example of a type you have already met, to see how the method works.

Example

Solve $u_n = 3u_{n-1} - 2u_{n-2} = 0$, $n \geq 2$, given $u_0 = 2$, $u_1 = 3$.

Solution

Let
$$G(x) = u_0 + u_1 x + u_2 x^2 + \ldots$$
$$= 2 + 3x + u_2 x^2 + \ldots \qquad (1)$$

Now from the original difference equation

$$u_2 = 3u_1 - 2u_0$$
$$u_3 = 3u_2 - 2u_1$$
$$u_4 = 3u_3 - 2u_2, \text{ etc.}$$

Substituting for u_2, u_3, u_4, ... into equation (1) gives

$$G(x) = 2 + 3x + (3u_1 - 2u_0)x^2 + (3u_2 - 2u_1)x^3 + \ldots$$

$$= 2 + 3x + \left(3u_1 x^2 + 3u_2 x^3 + 3u_3 x^4 + \ldots\right)$$
$$- \left(2u_0 x^2 + 2u_1 x^3 + 2u_2 x^4 + \ldots\right)$$

$$= 2 + 3x + 3x\left(u_1 x + u_2 x^2 + u_3 x^3 + \ldots\right)$$
$$- 2x^2\left(u_0 + u_1 x + u_2 x^2 + \ldots\right)$$

$$= 2 + 3x + 3x\left(G(x) - u_0\right) - 2x^2 G(x)$$

$$= 2 + 3x + 3x\left(G(x) - 2\right) - 2x^2 G(x)$$

$$G(x) = 2 - 3x + 3x\,G(x) - 2x^2 G(x).$$

Rearranging so that $G(x)$ is the subject gives

$$G(x) = \frac{2 - 3x}{1 - 3x + 2x^2}.$$

Note that $1 - 3x + 2x^2$ is similar to the auxilliary equation you met previously but **not** the same.

Factorising the denominator gives

$$G(x) = \frac{2 - 3x}{(1 - 2x)(1 - x)}. \qquad (2)$$

You now use partial fractions in order to write $G(x)$ as a sum of two fractions. There are a number of ways of doing this which you should have met in your pure mathematics core studies.

So $\qquad G(x) = \dfrac{1}{1 - 2x} + \dfrac{1}{1 - x}.$

Now both parts of $G(x)$ can be expanded using the binomial theorem

$$(1 - 2x)^{-1} = \left(1 + 2x + (2x)^2 + (2x)^3 + \ldots\right)$$

and $\qquad (1 - x)^{-1} = \left(1 + x + x^2 + x^3 + \ldots\right).$

This gives $\quad G(x) = \left(1 + 2x + (2x)^2 + \ldots\right) + \left(1 + x + x^2 + \ldots\right)$

$$= (1 + 1) + (2x + x) + \left(2^2 x^2 + x^2\right)$$

$$+ \left(2^3 x^3 + x^x\right) + \ldots$$

$$= 2 + 3x + \left(2^2 + 1\right)x^2 + \left(2^3 + 1\right)x^3 + \ldots$$

As you can see, the nth term of $G(x)$ is $\left(2^n + 1\right)x^n$ and the coefficient of x^n is simply u_n - the solution to the difference equation.

So $\qquad u_n = 2^n + 1.$

Exercise 11E

1. Find the generating function associated with these difference equations and sequences.

   (a) $u_n = 2u_{n-1} + 8u_{n-2}$, given $u_0 = 0$, $u_1 = 1$, $n \geq 2$.

   (b) $u_n + u_{n-1} - 3u_{n-2} = 0$, given $u_0 = 2$, $u_1 = 5$, $n \geq 2$.

   (c) $u_n = 4u_{n-2}$, given $u_0 = 1$, $u_1 = 3$, $n \geq 2$.

   (d) 1, 2, 4, 8, 16, ...

2. Write these expressions as partial fractions :

   (a) $\dfrac{3x-5}{(x-3)(x+1)}$ (b) $\dfrac{1}{(2x-5)(x-2)}$

   (c) $\dfrac{x+21}{x^2-9}$

3. Write these expressions as power series in x, giving the nth term of each series :

   (a) $\dfrac{1}{1-x}$ (b) $\dfrac{1}{1-2x}$ (c) $\dfrac{1}{1+3x}$

   (d) $\dfrac{1}{(1-x)^2}$ (e) $\dfrac{3}{(1+2x)^2}$

4. Solve these difference equations by using generating functions :

   (a) $u_n - 3u_{n-1} + 4u_{n-2} = 0$, given $u_0 = 0$, $u_1 = 20$, $n \geq 2$.

   (b) $u_n = 4u_{n-1}$, given $u_0 = 3$, $n \geq 1$.

5. Find the generating function of the Fibonacci sequence.

6. Find the particular solution of the difference equation $u_{n+2} = 9u_n$, given $u_0 = 5$, $u_1 = -3$, $n \geq 0$.

11.6 Extending the method

This final section shows how to solve non-homogeneous equations of the form :

$$\boxed{u_n + au_{n-1} + bu_{n-2} = f(n)} \quad (a, b \text{ constants}) \quad (1)$$

using the generating function method. The techniques which follow will also work for first order equations (where $b = 0$).

Example

Solve $u_n + u_{n-1} - 6u_{n-2} = n$, $n \geq 2$, given $u_0 = 0$, $u_1 = 2$.

Solution

Let the generating function for the equation be

$$G(x) = u_0 + u_1 x + u_2 x^2 + \dots.$$

Now work out $(1 + x - 6x^2)G(x)$.

The term $(1 + x - 6x^2)$ comes from the coefficients of u_n, u_{n-1} and u_{n-2} in the equation.

Now $\left(1+x-6x^2\right)G(x) = \left(1+x-6x^2\right)\left(u_0 + u_1 x + u_2 x^2 + ...\right)$

$$= u_0 + \left(u_1 + u_2\right)x + \left(u_2 + u_1 - 6u_0\right)x^2 + ...$$

$$= 0 + 2x + 2x^2 + 3x^3 + 4x^4 + ... \qquad (2)$$

In the last step values have been substituted for $u_n + u_{n-1} - 6u_{n-2}$. For example $u_3 + u_2 - 6u_1 = 3$.

The process now depends on your being able to sum the right-hand side of equation (2). The difficulty depends on the complexity of $f(n)$.

You should recognise (from Exercise 11E , Question 3) that

$$1 + 2x + 3x^2 + 4x^3 + ... = \frac{1}{(1-x)^2}.$$

So $\left(1+x-6x^2\right)G(x) = x\left(2 + 2x + 3x^2 + 4x^3 + ...\right)$

$$= 2x + x\left(2x + 3x^2 + 4x^3 + ...\right)$$

$$= 2x + x\left(\frac{1}{(1-x)^2} - 1\right)$$

$$= 2x + \frac{x}{(1-x)^2} - x$$

$$= x + \frac{x}{(1-x)^2}.$$

$$\Rightarrow (1+3x)(1-2x)G(x) = \frac{x(1-x)^2 + x}{(1-x)^2}$$

$$= \frac{x^3 - 2x^2 + 2x}{(1-x)^2}$$

$$\Rightarrow \qquad G(x) = \frac{x^3 - 2x^2 + 2x}{(1+3x)(1-2x)(1-x)^2}.$$

This result has to be reduced to partial fractions.

Let $\quad G(x) \equiv \dfrac{A}{1+3x} + \dfrac{B}{1-2x} + \dfrac{C}{(1-x)^2} + \dfrac{D}{1-x}$

$$\equiv \dfrac{x^3 - 2x^2 + 2x}{(1-3x)(1-2x)(1-x)^2},$$

then solving in the usual way gives $A = -\frac{5}{16}$, $B = 1$, $C = -\frac{1}{4}$, $D = -\frac{7}{16}$.

Thus $\quad G(x) = \dfrac{-5}{16(1+3x)} + \dfrac{1}{(1-2x)} - \dfrac{1}{4(1-x)^2} - \dfrac{7}{16(1-x)}$

$$= -\tfrac{5}{16}\left(1 - 3x + (-3x)^2 + ...\right) + \left(1 + 2x + (2x)^2 + ...\right)$$

$$-\tfrac{1}{4}\left(1 + 2x + 3x^2 + ...\right) - \tfrac{7}{16}\left(1 + x + x^2 + ...\right).$$

Picking out the coefficient of the nth term in each bracket gives

$$u_n = -\tfrac{5}{16}(-3)^n + 2^n - \tfrac{1}{4}(n+1) - \tfrac{7}{16}$$

$$= \tfrac{1}{16}\left[-5(-3)^n + 16 \times 2^n - 4(n-1) - 7\right].$$

As you can see, the form of u_n is still $Am_1^n + Bm_2^n$, but with the addition of a term of the form $Cn + D$. This additional term is particular to the function $f(n)$, which was n in this case, and to the values of u_0 and u_1.

Exercise 11F

1. Write the following as partial fractions.

   (a) $\dfrac{1}{(1-x)(1-2x)}$

   (b) $\dfrac{2x-3}{(2-x)(1+x)}$

   (c) $\dfrac{x^2+2}{(x+1)(1-2x)^2}$

2. Sum these series. Each is the result of expanding the expression of the form $(a+bx)^n$ using the Binomial Theorem (commonly $n = -1$ or -2).

   (a) $1 + x + x^2 + x^3 + ...$

   (b) $1 + 2x + 3x^2 + 4x^3 + 5x^4 + ...$

   (c) $-2 - 3x - 4x^2 - 5x^3 - 6x^4 - ..$

   (d) $x^2 + 2x^3 + 3x^4 + 4x^5 + ...$

   (e) $1 + 2x + (2x)^2 + (2x)^3 + (2x)^4 + ...$

   (f) $(5x)^2 + (5x)^3 + (5x)^4 + ...$

3. Expand as power series :

   (a) $(1-3x)^{-1}$ (b) $\dfrac{1}{(2-x)^2}$ (c) $\dfrac{x}{x-1}$.

   Give the nth term in each case.

4. Using the generating function method, solve

   (a) $u_n - 2u_{n-1} - 8u_{n-2} = 8$, $n \geq 2$, given $u_0 = 0$, $u_1 = 2$.

   (b) $u_n - 2u_{n-1} = 3^n$, $n \geq 1$, given $u_0 = 1$.

   (c) $u_n - u_{n-1} - 2u_{n-2} = n^2$, $n \geq 2$, given $u_0 = 0$, $u_1 = 1$.

11.7 Miscellaneous Exercises

1. Solve $u_n - 4u_{n-2} = 0$, $n \geq 3$, when $u_1 = 2$ and $u_2 = 20$.

2. Find the general solution of $u_n - 4u_{n-1} + 4u_{n-2} = 0$.

3. The life of a bee is quite amazing. There are basically three types of bee :

  queen   a fertile female

  worker   an infertile female

  drone   a fertile male.

  Eggs are either fertilised, resulting in queens and workers or unfertilised, resulting in drones.

  Trace back the ancestors of a drone. Find the numbers of ancestors back to the nth generation. The generation tree has been started for you below :

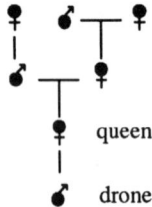

4. Find the nth term of these sequences :

  (a) 2, 5, 11, 23, ...

  (b) 2, 5, 12, 27, 58, ...

  (c) 1, 2, 6, 16, 44, ...

5. Solve the difference equation
$$u_n - u_{n-1} - 12u_{n-2} = 2^n, \ n \geq 2,$$
   if $u_0 = 0$, and $u_1 = 1$.

6. Write down the general solution of the model of the economy in Activity 4 when $A = \frac{2}{3}$, $B = 4$ and $N_1 = 1$, $N_2 = 2$.

7. A difference equation of the form
$$u_n + au_{n-1} + bu_{n-2} = k$$
   defines a sequence with its first five terms as 0, 2, 5, 5, 14. Find the nth term.

8. Find the smallest value of n for which u_n exceeds one million if $u_n = 10 + 3u_{n-1}$, $n \geq 1$, given $u_0 = 0$.

9. Find the solution of the difference equation
$$u_n = u_{n-2} + n, \ n \geq 2,$$
   given $u_1 = u_0 = 1$.

12 CRITICAL PATH ANALYSIS

Objectives

After studying this chapter you should

- be able to construct activity networks;
- be able to find earliest and latest starting times;
- be able to identify the critical path;
- be able to translate appropriate real problems into a suitable form for the use of critical path analysis.

12.0 Introduction

A complex project must be well planned, especially if a number of people are involved. It is the task of management to undertake the planning and to ensure that the various tasks required in the project are completed in time.

Operational researchers developed a method of scheduling complex projects shortly after the Second World War. It is sometimes called **network analysis**, but is more usually known as **critical path analysis** (**CPA**). Its virtue is that it can be used in a wide variety of projects, and was, for example, employed in such diverse projects as the Apollo moonshot, the development of Concorde, the Polaris missile project and the privatisation of the electricity and water boards. Essentially, CPA can be used for any multi-task complex project to ensure that the complete scheme is completed in the minimum time.

Although its real potential is for helping to schedule complex projects, we will illustrate the use of CPA by applying it to rather simpler problems. You will often be able to solve these problems without using CPA, but it is an understanding of the concepts involved in CPA which is being developed here.

12.1 Activity networks

In order to be able to use CPA, you first need to be able to form
what is called an **activity network**. This is essentially a way of
illustrating the given project data concerning the tasks to be
completed, how long each task takes and the constraints on the
order in which the tasks are to be completed. As an example,
consider the activities shown below for the construction of a
garage.

	activity	duration (in days)
A	prepare foundations	7
B	make and position door frame	2
C	lay drains, floor base and screed	15
D	install services and fittings	8
E	erect walls	10
F	plaster ceiling	2
G	erect roof	5
H	install door and windows	8
I	fit gutters and pipes	2
J	paint outside	3

Clearly, some of these activities cannot be started until other
activities have been completed. For example

activity G - erect roof

cannot begin until

activity E - erect walls

has been completed. The following table shows which activities
must precede which.

D must follow E

E must follow A and B

F must follow D and G

G must follow E

H must follow G

I must follow C and F

J must follow I.

We call these the **precedence relation**s.

All this information can be represented by the network shown below.

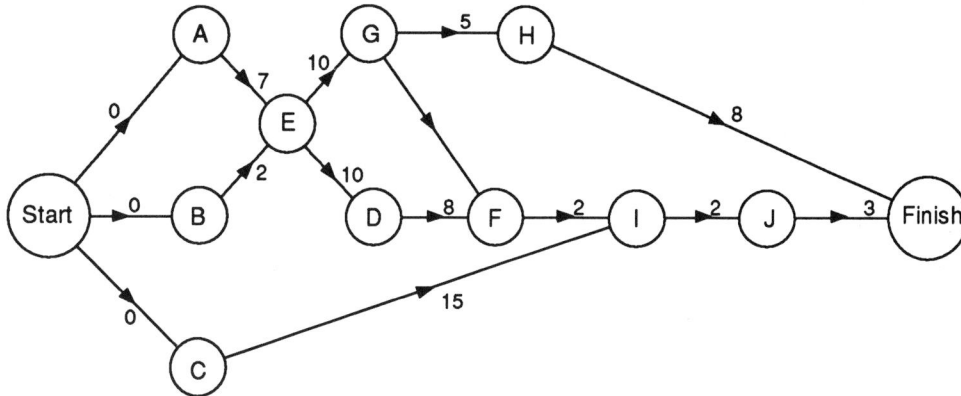

In this network

each **activity** is represented by a **vertex**;

joining vertex X to vertex Y shows that
activity X must be **completed** before Y can be started;

the **number** marked on each arc shows the **duration** of the
activity from which the arc starts.

Note the use of 'arc' here to mean a directed edge.
Sometimes we can easily form the activity network, but not
always, so we need to have a formal method. First try the
following activity.

Activity 1   Making a settee

A furniture maker is going to produce a new wooden framed settee
with cloth-covered foam cushions. These are the tasks that have to
be done by the furniture maker and his assistants and the times
they will take :

	activity	time in days
A	make wooden arms and legs	3
B	make wooden back	1
C	make wooden base	2
D	cut foam for back and base	1
E	make covers	3
F	fit covers	1
G	put everything together	1

Each activity can only be undertaken by one individual.

The following list gives the order in which the jobs must be done:

> B   must be after   C
>
> A   must be after   B and C
>
> D   must be after   B and C
>
> E   must be after   D
>
> F   must be after   E
>
> G   must be after   A, B, C, D, E and F

Construct an appropriate activity network to illustrate this information.

12.2 Algorithm for constructing activity networks

For simple problems it is often relatively easy to construct activity networks but, as the complete project becomes more complex, the need for a formal method of constructing activity networks increases. Such an algorithm is summarised below.

Start	Write down the original vertices and then a second copy of them alongside, as illustrated on the right. If activity Y must follow activity X draw an arc from original vertex Y to shadow vertex X. (In this way you construct a **bipartite graph**.)
Step 1	Make a list of all the original vertices which have **no** arcs incident to them.
Step 2	Delete all the vertices found in Step 1 and their corresponding shadow vertices and all arcs incident to these vertices.
Step 3	Repeat Steps 1 and 2 until all the vertices have been used.

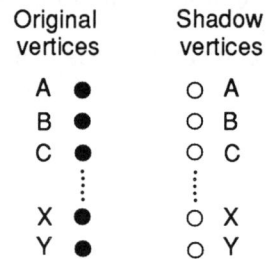

Original vertices

Shadow vertices

A ●   ○ A
B ●   ○ B
C ●   ○ C
⋮   ⋮
X ●   ○ X
Y ●   ○ Y

The use of this algorithm will be illustrated using the first case study, constructing a garage, from Section 12.1.

The precedence relations are:

D must follow E

E must follow A and B

F must follow D and G

G must follow E

H must follow G

I must follow C and F

J must follow I

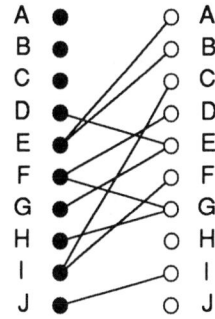

These are illustrated opposite.

Applying the algorithm until all vertices have been chosen is shown below.

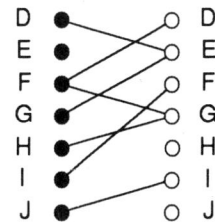

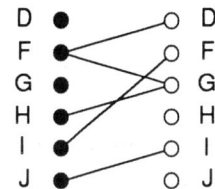

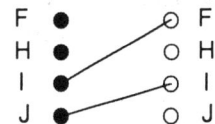

Step 1 - original vertices with no arcs

Step 2 - delete all arcs incident on A, B, C and redraw as shown

Step 3 - repeat iteration

A, B, C

Step 1 - original vertices with no arcs

Step 2 - delete all arcs incident on E and redraw as shown

Step 3 - repeat iteration

E

Step 1 - original vertices with no arcs

Step 2 - delete all arcs incident on D, G and redraw as shown

Step 3 - repeat iteration

D, G

Step 1 - original vertices with no arcs

Step 2 - delete all arcs incident on F, H and redraw as shown

Step 3 - repeat iteration

F, H

Step 1 - original vertices with no arcs

Step 2 - delete all arcs incident on I and redraw as shown

I

Step 3 - stop as all vertices have been chosen

So the vertices have been chosen in the following order:

A

		D	F		
B	E			I	J
		G	H		

C

The activity diagram as shown belowcan now be drawn.

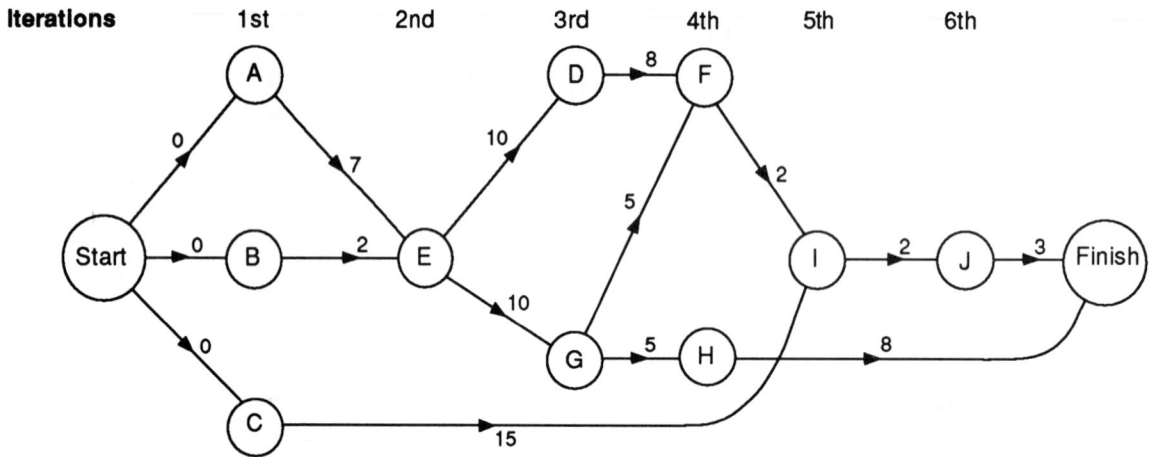

From the 'start' vertex, draw arcs to A, B and C, the first iteration vertices, putting zero on each arc. In the original bipartite graph the shadow vertex A was joined to the original vertes F - so join A to E. Similarly join B to G and C to I.

Indicate the duration of the activity on any arc coming **from** the vertex representing the activity.

Continue in this way and complete the activity network with a 'finish' vertex into which any free vertices lead, again indicating the duration of the activity on the arc.

Note that the duration of the activity is shown on every arc **coming** from the vertex representing the activity. (So, for example, arc ED and arc EG are both given 10.)

Exercise 12A

1. Use the algorithm to find the activity network for the problem in Activity 1.

2. Suppose you want to redecorate a room and put in new self-assembly units. These are the jobs that need to be done, together with the time each takes:

activity	time (in hrs)	preceded by
paint woodwork (A)	8	-
assemble units (B)	4	-
fit carpet (C)	5	hang wallpaper paint woodwork
hang wallpaper (D)	12	paint woodwork
hang curtains (E)	2	hang wallpaper paint woodwork

Complete an activity network for this problem.

3. The Spodleigh Bicycle Company is getting its assembly section ready for putting together as many bicycles as possible for the Christmas market. This diagram shows the basic components of a bicycle.

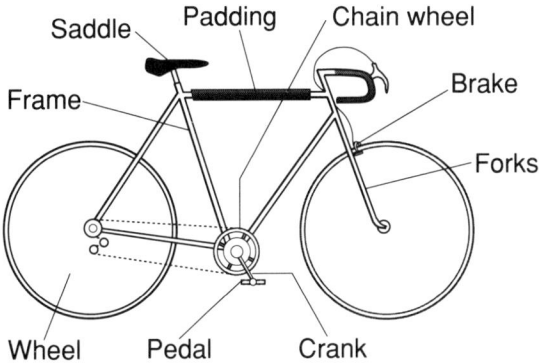

Putting together a bicycle is split up into small jobs which can be done by different people. These are:

activity	time (mins)
A preparation of the frame	9
B mounting and aligning the front wheel	7
C mounting and aligning the back wheel	7
D attaching the chain wheel to the crank	2
E attaching the chain wheel and crank to the frame	2
F mounting the right pedal	8
G mounting the left pedal	8
H final attachments such as saddle, chain, stickers, etc.	21

The following chart shows the order of doing the jobs.

B must be after A

C must be after A

D must be after A

E must be after D

F must be after D and E

G must be after D and E

H must be after A, B, C, D, E, F and G

Draw an activity network to show this information.

4. An extension is to be built to a sports hall. Details of the activities are given below.

activity	time (in days)
A lay foundations	7
B build walls	10
C lay drains and floor	15
D install fittings	8
E make and fit door frames	2
F erect roof	5
G plaster ceiling	2
H fit and paint doors and windows	8
I fit gutters and pipes	2
J paint outside	3

Some of these activities cannot be started until others have been completed:

B must be after C

C must be after A

D must be after B

E must be after C

F must be after D and E

G must be after F

H must be after G

I must be after F

J must be after H

Complete an activity network for this problem.

12.3 Critical path

You have seen how to construct an activity network. In this section you will see how this can be used to find the **critical path**. This will first involve finding the **earliest** possible start for each activity, by going **forwards** through the network. Secondly, the **latest** possible start time for each activity is found by going **backwards** through the network. Activities which have **equal** earliest and latest start time are on the **critical path**. The technique will be illustrated using the 'garage construction' problem from Sections 12.1 and 12.2.

The activity network for this problem is shown below, where sufficient space is made at each activity node to insert two numbers.

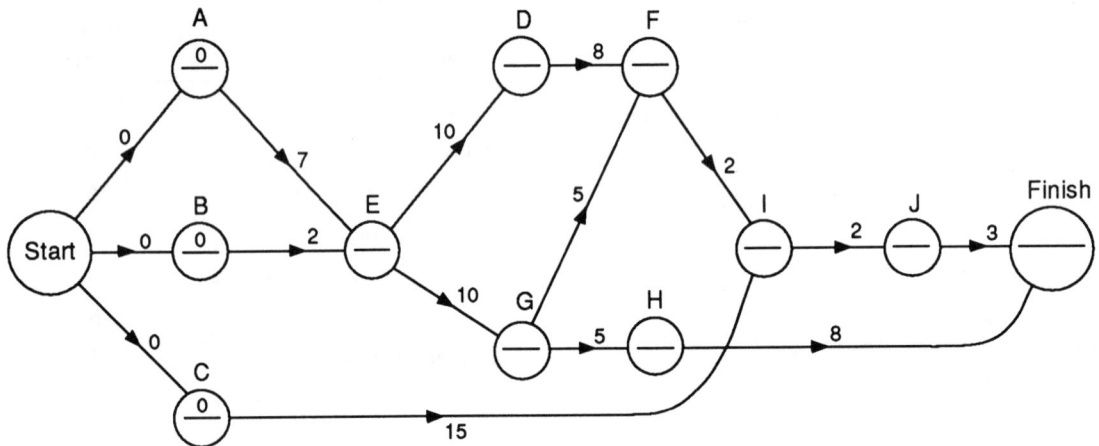

The numbers in the top half of each circle will indicate the earliest possible starting time. So, for activities A, B and C, the number zero is inserted.

Moving forward through the network, the activity E is reached next. Since both A and B have to be completed before E can be started, the earliest start time for E is 7. This is put into the top half of the circle at E. The earliest times at D and G are then both 17, and for H, 22. Since F cannot be started until both D and G are completed, its earliest start time is 25, and consequently, 27 for I. The earliest start time for J is then 29, which gives an earliest completion time of 32.

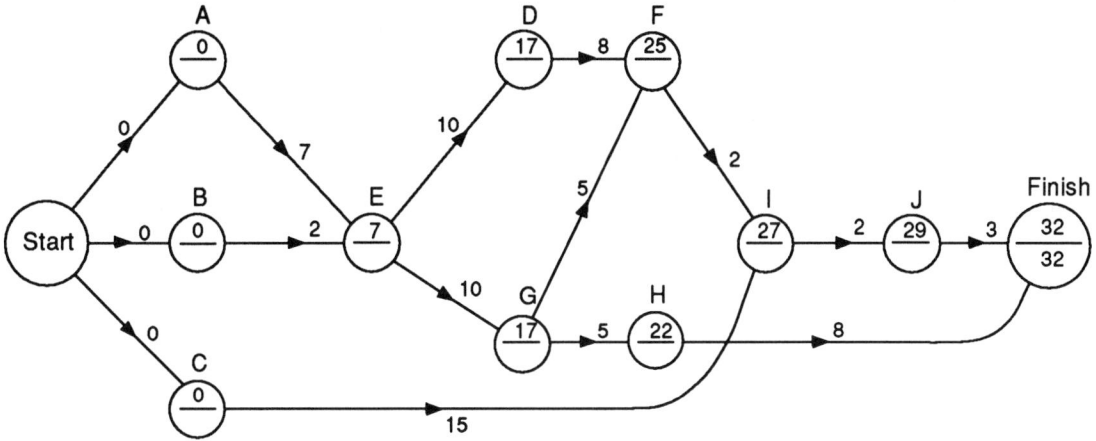

Since 32 is the earliest possible completion time, it is also assumed to be the completion time in order to find the latest possible start times. So 32 is also put in the lower half of the 'finish' circle. Now working backwards through the network, the latest start times for each activity are as follows:

J $32 - 3 = 29$

I $29 - 2 = 27$

F $27 - 2 = 25$

H $32 - 8 = 24$

D $25 - 8 = 17$

G   the minimum of $25 - 5 = 20$ and $24 - 5 = 19$

E   the minimum of $17 - 10 = 7$ and $19 - 10 = 9$

A $7 - 7 = 0$

B $7 - 2 = 5$

C $27 - 15 = 12$

This gives a completed network as shown below.

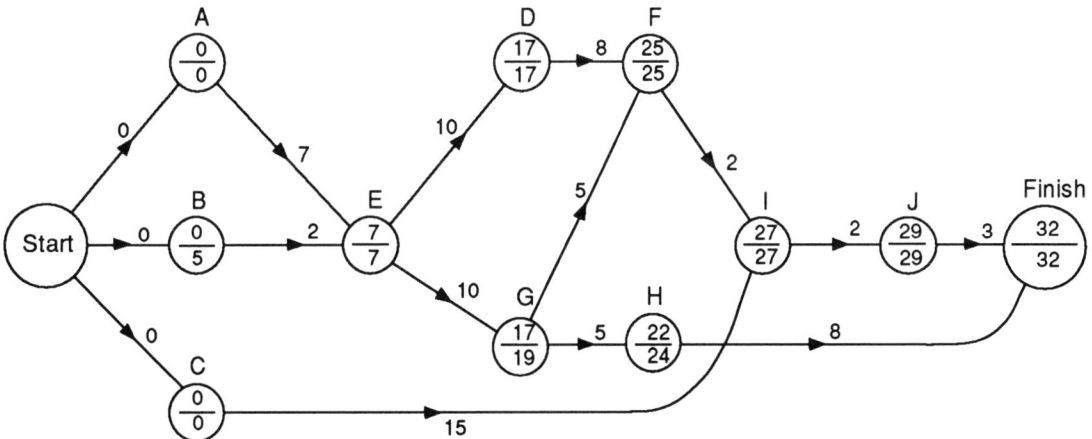

The vertices with equal earliest and latest starting times define the **critical path**. This is clearly seen to be

$$A \; E \; D \; F \; I \; J.$$

Another way of identifying the critical path is to define the

float time = latest start time − earliest start time.

The information for the activities can now be summarised in the table below.

activity	start times earliest	latest	float	
A	0	0	0	←
B	0	5	5	
C	0	12	12	
E	7	7	0	←
D	17	17	0	←
G	17	19	2	
F	25	25	0	←
H	22	24	2	
I	27	27	0	←
J	29	29	0	←

So now you know that if there are enough workers the job can be completed in 32 days. The activities on the critical path (i.e. those with zero float time) must be started punctually; for example, A must start immediately, E after 7 days, F after 25 days, etc. For activities with a non-zero float time there is scope for varying their start times; for example activity G can be started any time after 17, 18 or 19 days' work. Assuming that all the work is completed on time, you will see that this does indeed give a working schedule for the construction of the garage in the minimum time of 32 days. However it does mean, for example, that on the 18th day activities D and G will definitely be in progress and C may be as well. The solution could well be affected if there was a limit to the number of workers available, but you will consider that sort of problem in the next chapter.

Is a critical path always uniquely defined?

Activity 2　　Bicycle construction

From the activity network for Question 3 in Exercise 12A find the critical path and the possible start times for all the activities in order to complete the job in the shortest possible time;

Exercise 12B

1. Find the critical paths for each of the activity networks shown below.

(a)

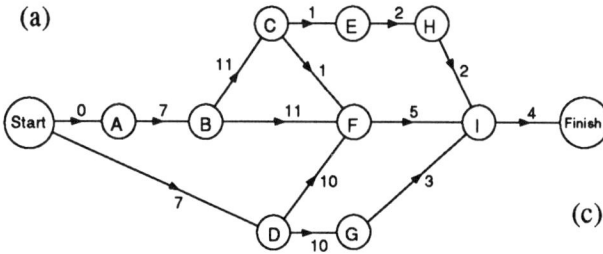

(b)

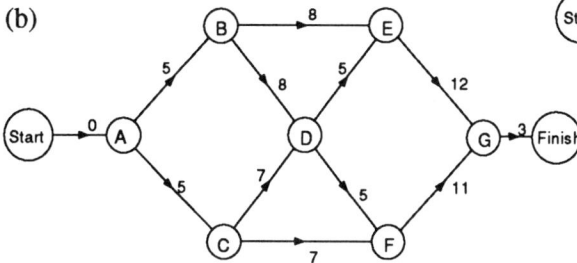

(c)

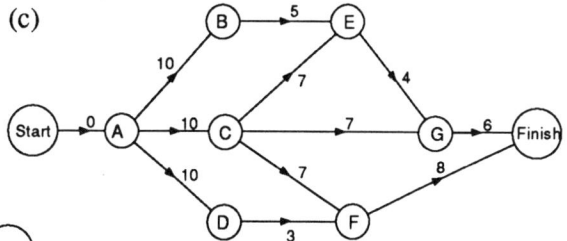

2. Find the critical path for the activity network in Question 4, Exercise 12A.

3. Your local school decides to put on a musical. These are the many jobs to be done by the organising committee, and the times they take:

A	make the costumes	6 weeks
B	rehearsals	12 weeks
C	get posters and tickets printed	3 weeks
D	get programmes printed	3 weeks
E	make scenery and props	7 weeks
F	get rights to perform the musical	2 weeks
G	choose cast	1 week
H	hire hall	1 week
I	arrange refreshments	1 week
J	organise make-up	1 week
K	decide on musical	1 week
L	organise lighting	1 week
M	dress rehearsals	2 days
N	invite local radio/press	1 day
P	choose stage hands	1 day
Q	choose programme sellers	1 day
R	choose performance dates	$\frac{1}{2}$ day
S	arrange seating	$\frac{1}{2}$ day
T	sell tickets	last 4 weeks
V	display posters	last 3 weeks

(a) Decide on the precedence relationships.

(b) Construct the activity network.

(c) Find the critical path and minimum completion time.

12.4 Miscellaneous Exercises

1. Consider the following activity network, in
which the vertices represent activities and the
numbers next to the arcs represent time in days.

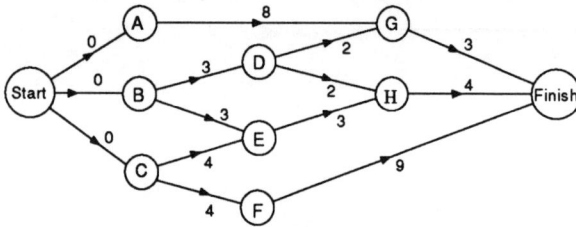

(a) Assuming that an unlimited number of
workers is available, write down:

(i) the minimum completion time of the
project if an unlimited number of
workers is available;

(ii)the corresponding critical path.

(b) Find the float time of activity E.

2. A project consists of ten activities, A-J. The
duration (in days) of each activity, and the
activities preceding each of them, are as follows:

activity	duration	preceding activities
A	10	-
B	4	-
C	8	B
D	6	C
E	8	I
F	5	-
G	10	A, D
H	2	G
I	4	-
J	10	D, F, I

Using the algorithms in Section 12.2,

(a) construct an activity network for this project;

(b) find a critical path in this activity network;

(c) find the latest starting time for each activity.

3. A project consists of eight activities whose
durations are as follows:

activity	A	B	C	D	E	F	G	H
duration	4	4	3	5	4	1	6	5

The precedence relations are as follows:

B must follow A

D must follow A and C

F must follow C and E

G must follow C and E

H must follow B and D.

(a) Draw an activity network in which the
activities are represented by vertices.

(b) Find a critical path by inspection, and write
down the earliest and latest starting times for
each activity.

4. The eleven activities A to K which make up a
project are subject to the following precedence
relations.

preceding activities	activity	duration
C, F, J	A	7
E	B	6
-	C	9
B, H	D	7
C, J	E	3
-	F	8
A, I	G	4
J	H	9
E, F	I	9
-	J	7
B, H, I	K	5

(a) Construct an activity network for the project.

(b) Find:

(i) the earliest starting time of each activity
in the network;

(ii) the latest starting time of each activity.

(c) Calculate the float of each activity, and
hence determine the critical path.

5. The activities needed to replace a broken window pane are given below.

	activity	duration (in mins)	preceding activities
A	order glass	10	-
B	collect glass	30	A
C	remove broken pane	15	B, D
D	buy putty	20	-
E	put putty in frame	3	C
F	put in new pane	2	E
G	putty outside and smooth	10	F
H	sweep up broken glass	5	C
I	clean up	5	all

(a) Construct an activity network.

(b) What is the minimum time to complete the replacement?

(c) What is the critical path?

6. Write the major activities, duration time and precedence relationship for a real life project with which you are involved. Use the methods in this chapter to find the critical path for your project.

13 SCHEDULING

Objectives

After studying this chapter you should

- be able to apply a scheduling algorithm to Critical Path Analysis problems;
- appreciate that this does not always produce the optimum solution;
- be able to design methods for solving packing problems;
- be able to use the branch-and-bound method for solving the knapsack problem.

13.0 Introduction

In the previous chapters it was possible to find the critical path for complex planning problems, but no consideration was given to how many workers would be available to undertake the activities, or indeed to how many workers would be needed for each activity.

You will see how scheduling methods can be applied to Critical Path Analysis problems but, importantly, it will be shown that such methods do not necessarily give the **optimal** solution every time.

You will also see how this scheduling problem is related to bin-filling problems, and to a similar problem called the knapsack problem in which the items carried not only have particular **weight**, but also have an appropriate **value**.

Activity 1

Find the critical path for the activity network shown below.

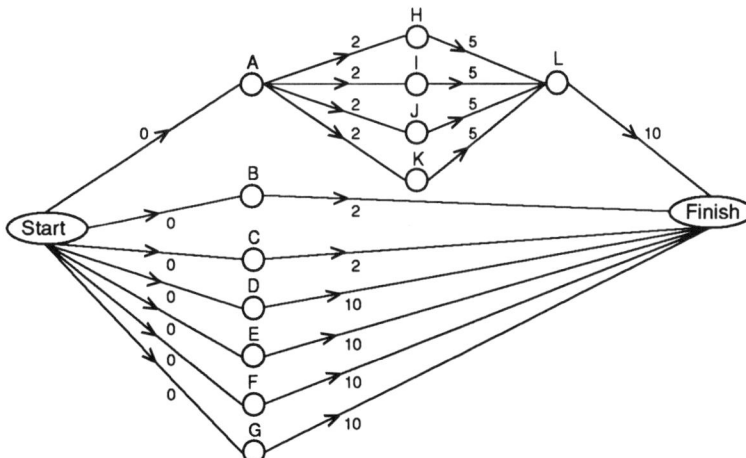

Suppose that each activity can be undertaken by a single worker, and that there are just 2 workers available. Also, that once an activity has been started by a worker, it must be completed by that same worker with no stoppages.

Design a schedule for these two workers so that the complete project is completed in the **minimum** time possible.

Does the minimum completion time depend on the number of workers available?

13.1 Scheduling

As in Activity 1, the following **operating rules** will be assumed:

1.	Each activity requires only **one** worker.
2.	No worker may be idle if there is an activity that can be started.
3.	Once a worker starts an activity, it must be continued by that worker until it is completed.

The **objective** will be to:

'Complete the project as soon as possible with the available number of workers.'

The main example from Chapter 12, which related to the construction of a garage, will be used to illustrate the problem.

The activity network, earliest and latest starting times, and the critical path (bold line), are shown below.

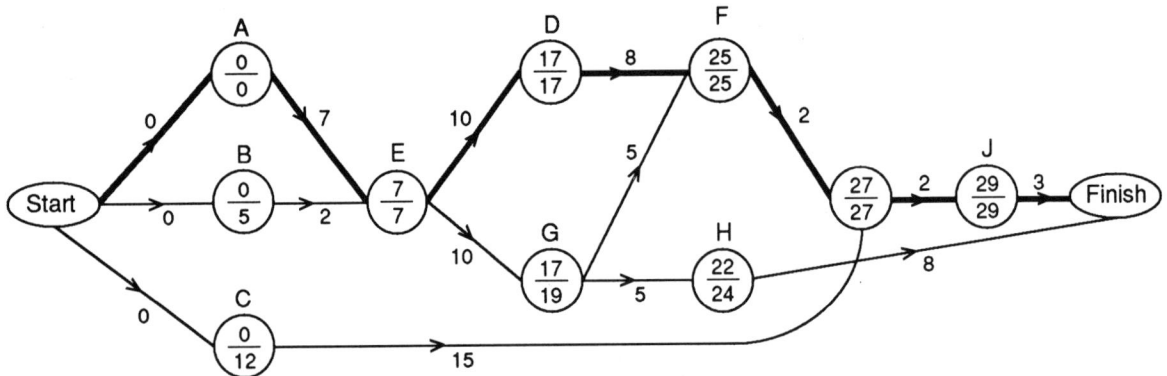

Suppose there are two workers available for the complete project. What is needed is a set procedure in order to decide who does what.

Can you think what would be a suitable procedure for allocating workers to activities?

The method that will be adopted can be summarised as follows.

> At any stage, when a worker becomes free, consider all the activities which have not yet been started but which can now be started. Assign to the worker the most 'critical' one of these (i.e. the one whose latest starting time is the smallest). If there are no activities which can be started at this stage you may have to wait until the worker can be assigned a job.

Using this as a basis for decisions, the solution shown opposite is obtained.

Note that worker 1 completes all the activities on the critical path, though, at time $t = 17$, workers 1 and 2 could have swopped over.

Since the whole project is completed in time 32 days, which you already know to be the minimum completion time, you can be assured that this method has produced an optimum solution. However, this is not always the case, as you will see in the next example.

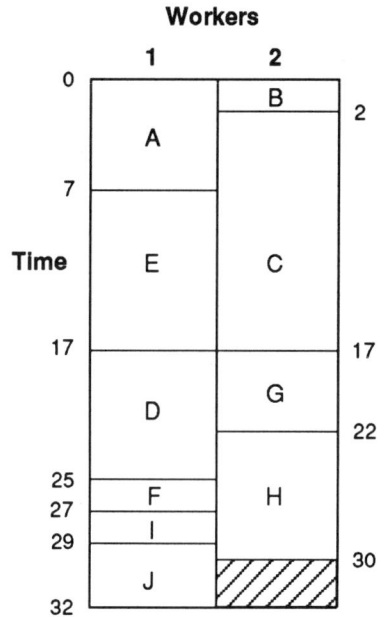

Example

The problem in Activity 1 has the following activity network and critical path. Use the method above to schedule four workers for this project.

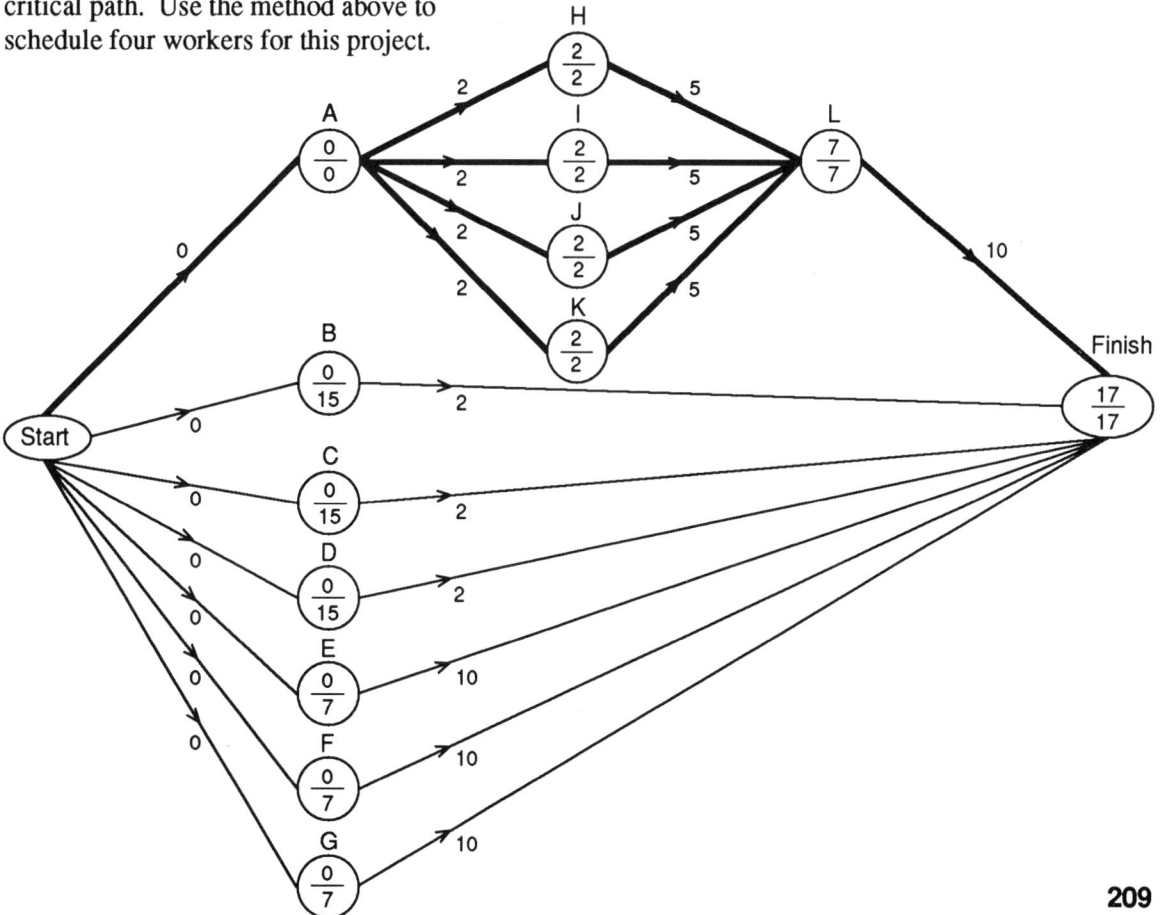

Solution

Workers

Applying the method as before results in the schedule shown in the first diagram opposite. This gives a completion time of $t = 25$.

Is this solution optimal?

It should not take too long to find a schedule which completes the project in time 17. A possible solution is shown in the second diagram.

As no worker is ever idle, and they all finish at time 17, this must be an optimum solution - you cannot do better! So the algorithm does not always produce the optimum solution. Currently no procedure exists which always guarantees to give the optimum scheduling solution, except the method of exhaustion where every possible schedule is tested.

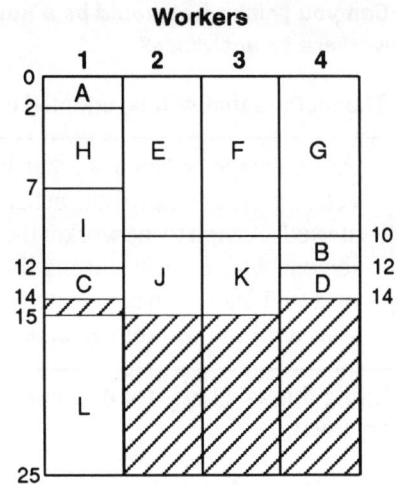

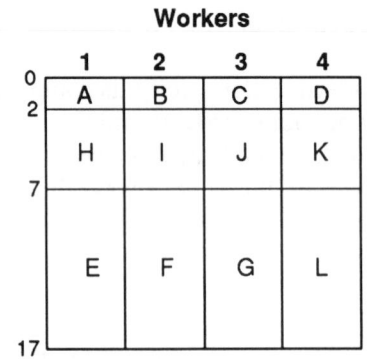

Activity 2

A possible revision of the method is to :

> **Evaluate** for each activity, the sum of the earliest and latest starting times, and **rank** the activities in ascending order according to this sum.

Then activities are assigned according to this ranking, taking the precedence relations into account. Use this method to schedule the project above, again using four workers. Does it produce the optimal solution?

Activity 3

Design your own method of scheduling. Try it out on the two examples above.

Exercise 13A

1. Use the first scheduling method to find a solution to Question 3 in Section 12.4, using two workers. Does this produce an optimum schedule?

2. Find a schedule for the problem given in Question 3, Exercise 12A, using 2 workers and the two scheduling algorithms given in this section.

   Does either of these methods provide an optimal solution?

3. A possible modification to the method in this section is as follows :

   'Evaluate for each activity the **product** of the earliest and latest starting times, and rank the activities in ascending order according to these numbers. Assign activities using this ranking, taking the precedence relations into account.'

   Use this method to find possible schedules for the garage construction problem in this section. Does this method always give an optimum solution?

13.2 Bin packing

In the previous section, you saw how to schedule activities for a given number of workers in order to complete the project in minimum time. In this section, the problem is turned round and essentially asks for the minimum number of workers required to complete the project within a given time. It will be assumed that there are no precedence relations. The difficulties will be illustrated in the following problem.

A project consists of the following activities (with no precedence relations):

Activity	A	B	C	D	E	F	G	H	I	J	K
Duration (in days)	8	7	4	9	6	9	5	5	6	7	8

What is the minimum number of workers required to complete the project in 15 days?

Find a lower bound to the minimum number of workers needed.

Activity 4

Show that there is a solution to this problem which uses only five workers.

You should obtain a solution without too much difficulty. It is more difficult, however, to find an **algorithm** to solve such problems. The problem considered here is one involving **bin packing**. If you replace workers by bins, each having a maximum capacity of 15 units, the problem is to use the minimum number of bins.

Think about how a precise method could be designed to solve problems of this type.

One possible method is known as **first-fit packing** :

> Number the bins, then always place the next item in the lowest numbered bin which can take that item.

Applying this method to the problem above gives the following solution.

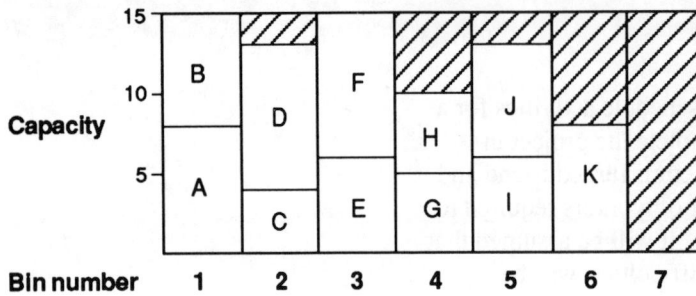

Using this method, six bins are needed, but you should have found a solution in Activity 4 which needs just five bins. So this method does not necessarily produce the optimum solution.

How can the first-fit method be improved?

Looking at the way the method works, it seems likely that it might be improved by just reordering the items into decreasing order of size, so that the items of largest size are packed first. Then you have the **first-fit decreasing method** :

1. Reorder the items in decreasing order of size.

2. Apply the first fit procedure to the reordered list.

You will see how this works using the same problem as above. First reorder the activities in decreasing size.

Activity	D	F	A	K	B	J	E	I	G	H	C
Duration (in days)	9	9	8	8	7	7	6	6	5	5	4

and then apply the method to give the solution below.

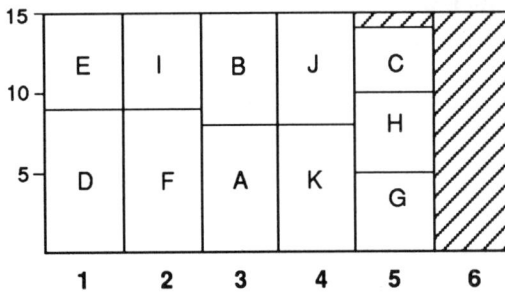

This clearly gives an optimal solution.

Will this method always give an optimal solution?

Bin-packing problems occur in a variety of contexts. As you have already seen, one context is that of determining the minimum number of workers to complete a project in a specified time period. Other examples occur in :

Plumbing   in which it is required to minimise the number of pipes of standard length required to cut a specified number of different lengths of pipe.

Advertising on television, in which case the bins are the standard length breaks between programmes, with the problem of trying to pack a specified list of adverts into the smallest number of breaks.

Example

A builder has piping of standard length 12 metres.

The following sections of various lengths are required

Section	A	B	C	D	E	F	G	H	I	J	K	L
Length (in metres)	2	2	3	3	3	3	4	4	4	6	7	7

Find a way of cutting these sections from the standard 12 m lengths so that a minimum number of lengths is used. Use

(a)   first-fit method,

(b)   first-fit decreasing method,

(c)   trial and error,

to find a solution.

Solution

(a) First-fit method

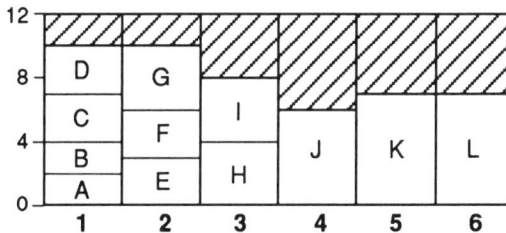

(b) First-fit decreasing method

Section	L	K	J	I	H	G	F	E	D	C	B	A
Length (in metres)	7	7	6	4	4	4	3	3	3	3	2	2

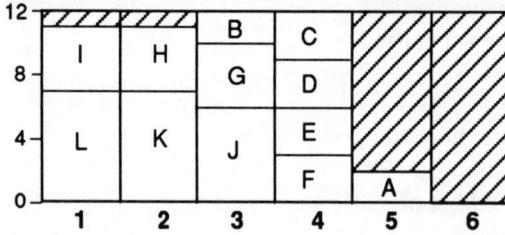

(c) Trial and Error

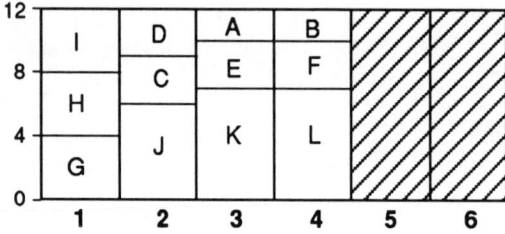

Note that even the first-fit decreasing method does not necessarily give the optimum solution, as shown above. An indication of the number of bins required can be obtained by evaluating

$$\frac{\text{sum of all sizes}}{\text{bin size}}$$

and noting the smallest integer that is greater than (or equal to) this number. This integer is a **lower bound** to the number of bins required. However you cannot always obtain a solution with this number.

Example

Find an optimum solution for fitting items of size

$$7, \; 6, \; 6, \; 6, \; 4, \; 3$$

into bins of size 11.

Solution

Noting that

$$\frac{\text{sum of all sizes}}{\text{bin size}} = \frac{7+6+6+6+4+3}{11} = \frac{32}{11} = 2\tfrac{10}{11},$$

it can be see that three is a lower bound to the number of bins required. But it is clear that four bins will in fact be needed and that no solution exists using just three bins.

A possible solution is given opposite.

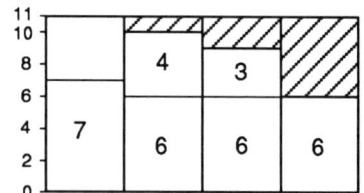

Exercise 13B

1. A project consists of eight activities whose durations are as follows. There are no precedence relations.

Activity	A	B	C	D	E	F	G	H
Duration (hours)	1	2	3	4	4	3	2	1

   It is required to find the minimum number of workers needed to finish the project in 5 hours.

   Find the answers to this problem given by

   (a) the first-fit packing method;

   (b) the first-fit decreasing method.

2. A plumber uses pipes of standard length 10 m and wishes to cut out the following lengths

Length (m)	10	9	8	7	6	5	4	3	2	1
Number	0	0	2	3	1	1	0	2	3	0

   Use the first-fit decreasing method to find how many standard lengths are needed to meet this order. Does this method give an optimum solution? If not, find an optimum solution.

3. Determine the minimum number of sheets of metal required, 10 m by 10 m, to meet the following order, and how they should be cut. (Assume no wastage in cutting.)

Size	Number
$3 \times 1\,m^2$	60
$4 \times 2\,m^2$	49
$7 \times 5\,m^2$	12

   Develop a **general** method of solving 2-dimensional packing problems of this type.

*13.3 Knapsack problem

For the bin-filling problem, the aim was to pack items of different sizes into a minimum number of bins. Now suppose that there is just **one** bin, but that each item has a value associated with it. Thus the question is what items should be packed in order to maximise the total value of the items packed. The problem is known as the **knapsack** problem as it can be interpreted in terms of a hiker who can only carry a certain total weight in his/her knapsack (rucksack). The hiker has a number of items that he/she would like to take, each of which has a particular value The problem is to decide which items should be packed so that the total value is a maximum, subject to the weight restriction.

This type of problem will be solved using a technique called the **branch and bound method**. How it works will be shown using the following particular problem.

Suppose a traveller wishes to buy some books for his journey. He estimates the time it will take to read each of five books and notes the cost of each one :

Book	A	B	C	D	E
Cost (£)	4	6	3	2	5
Reading time (hours)	5	9	4	4	4

Which of these books should he buy to maximise his total reading time without spending more than £8?

Activity 5

By trial and error, find the solution to the traveller's problem.

As you have probably seen, with just a few items it is easy enough to find the optimum solution. However, in the example above, if there was a choice of, say, 10 or 15 books, the problem of finding the optimum solution would now be far more complex.

The first step in the branch and bound method is to list the items in decreasing order of reading time per unit cost.

Item	A	B	C	D	E	
Cost	4	6	3	2	5	('weight')
Reading time	5	9	4	4	4	('value')
Reading time per unit cost	1.25	1.5	1.33	2	0.8	('value per unit weight')

Reordering,

Number	1	2	3	4	5
Item	D	B	C	A	E
Cost (w)	2	6	3	4	5
Value (v)	4	9	4	5	4
Value per unit cost	2	1.5	1.33	1.25	0.8

A solution **vector** of the form $(x_1, x_2, x_3, x_4, x_5)$ will be used to denote a possible solution where

$$x_i = \begin{cases} 1 & \text{if item } i \text{ is bought} \\ 0 & \text{if item } i \text{ is not bought} \end{cases}$$

So, for example,

$$\mathbf{x} = (0, 0, 1, 0, 1)$$

means buying books C and E, which have total cost £8 and total value $4 + 4 = 8$ hours of reading time.

The method uses a branching method to search for the optimal solution.

For example let us look at the possible branches from $(1, 0, 0, 0, 0)$.

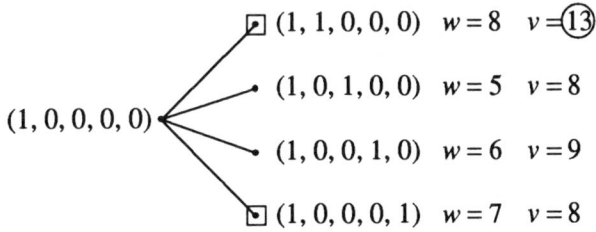

$(1, 0, 0, 0, 0)$ branches to:
- ▣ $(1, 1, 0, 0, 0)$ $w = 8$ $v = ⑬$
- • $(1, 0, 1, 0, 0)$ $w = 5$ $v = 8$
- • $(1, 0, 0, 1, 0)$ $w = 6$ $v = 9$
- ▣ $(1, 0, 0, 0, 1)$ $w = 7$ $v = 8$

The 'square' shows that there can be no further branching from this point. For example, there is a square by $(1, 1, 0, 0, 0)$ because the total allowed cost (weight) of 8 has been reached. Also we always add additional 1s to the right of the last 1 so that, for example, you could branch from $(1, 0, 0, 1, 0)$ to $(1, 0, 0, 1, 1)$ but you would never consider branching from $(1, 0, 0, 1, 0)$ to $(1, 0, 1, 1, 0)$. So when, in the example above, x_5 is 1, no further branching is possible : that accounts for the square by $(1, 0, 0, 0, 1)$. Note that the best solution (i.e. maximum v) at this stage is $v = 13$.

You start the full process at the **null** solution $(0, 0, 0, 0, 0)$, now written more simply as $0\,0\,0\,0\,0$, and then keep repeating the process as outlined below.

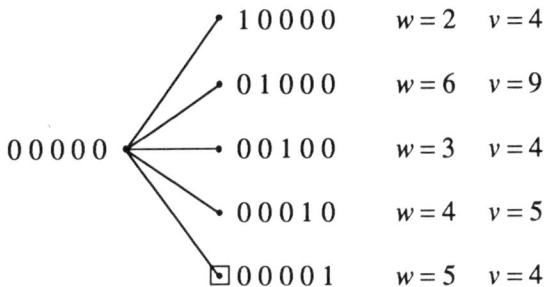

$0\,0\,0\,0\,0$ branches to:
- $1\,0\,0\,0\,0$ $w = 2$ $v = 4$
- $0\,1\,0\,0\,0$ $w = 6$ $v = 9$
- $0\,0\,1\,0\,0$ $w = 3$ $v = 4$
- $0\,0\,0\,1\,0$ $w = 4$ $v = 5$
- ▣ $0\,0\,0\,0\,1$ $w = 5$ $v = 4$

You can now branch from any of the four vertices which are not squared : for example, the branches from $1\,0\,0\,0\,0$ are as shown above and the branches from $0\,1\,0\,0\,0$ are as shown below :

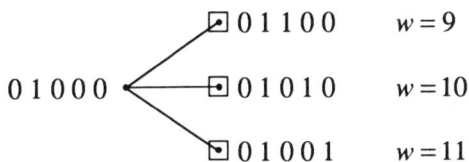

$0\,1\,0\,0\,0$ branches to:
- ▣ $0\,1\,1\,0\,0$ $w = 9$
- ⊡ $0\,1\,0\,1\,0$ $w = 10$
- ▣ $0\,1\,0\,0\,1$ $w = 11$

Continue in this way to give one single diagram, as shown on the next page :

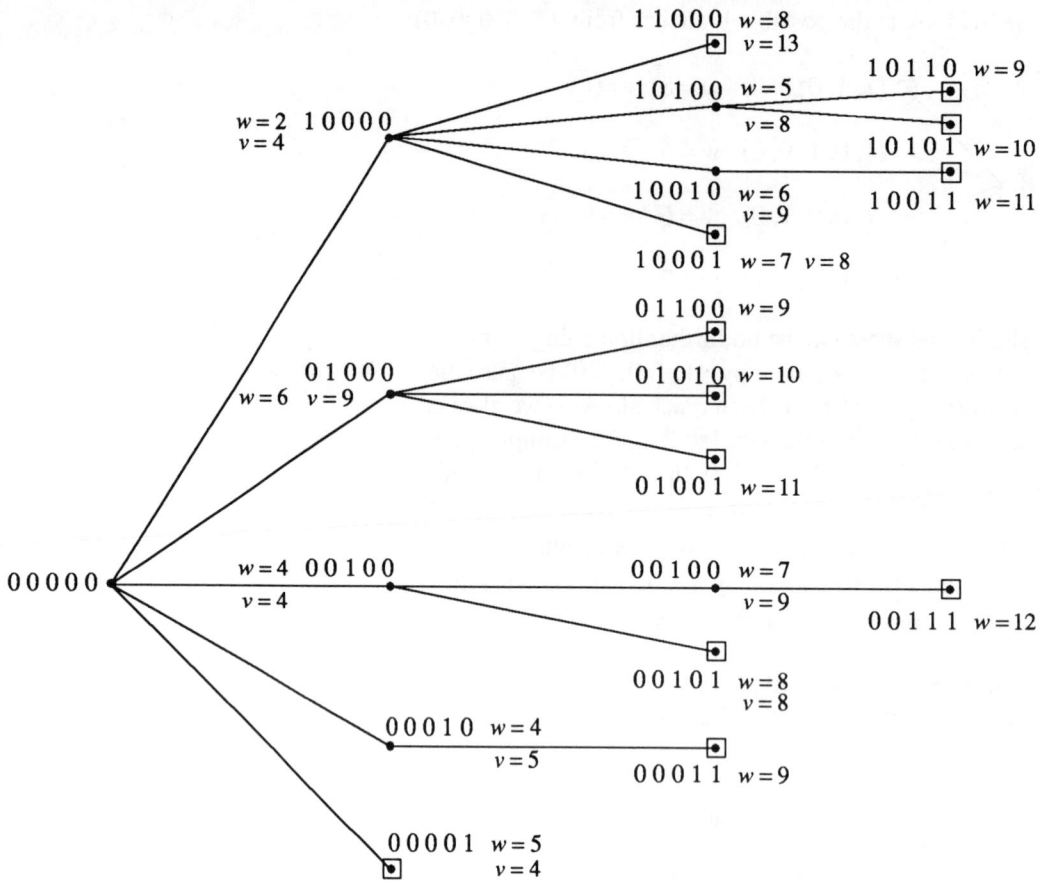

Then from amongst the squared vertices with $w \leq 8$, you find the one with the highest v, namely $1\,1\,0\,0\,0$ in the this case. That means that the traveller should take books D and B, giving a total reading time of 13 hours.

Exercise 13C

1. Suppose that a hiker can pack up to 9 kg of items and that the following items are available to take. The value of each item is also specified.

Item	A	B	C	D	E
Weight (kg)	3	8	6	4	2
Value	2	12	9	3	5

Use the branch and bound method to find the items that can be taken which give a maximum total value.

2. The manager of a firm which has installed a small computer system has received requests from four potential users. The computer will be run for up to 24 hours per day but can cope with only one user at a time. The manager estimates that the number of hours of computer time required by each user per day, and the consequent likely income for the firm, are as follows:

User	A	B	C	D
Use (hours/day)	8	12	13	4
Income (£1000/year)	72	102	143	38

Use the branch and bound method to determine which users should be allocated time on the computer system so as to maximise the total income.

3. A machine in a factory can be used to make any one of five items, A, B, C, D and E. The time taken to produce each item, and the value of each item, are shown in the following table.

Item	A	B	C	D	E
Production time (in days)	3	7	2	4	4
Value	3	14	3	7	8

If the machine is available for only 10 days, use the branch and bound method to determine which of the items should be produced so that the total value is as large as possible.

13.4 Miscellaneous Exercises

1. A project consists of ten activities A-J with the following durations (in hours). There are no precedence relations.

Activity	A	B	C	D	E	F	G	H	I	J
Duration	2	3	4	5	6	7	8	9	10	11

   (a) Find the minimum number of workers needed to complete this project in 16 hours.

   (b) Use (i) the first-fit packing method,

   (ii) the first-fit decreasing method.

   Does either of these methods produce an optimum solution?

2. A hiker wishes to take with her a number of items. Their weights and values are given in the table below.

Item	A	B	C	D	E
Weight (kg)	5	4	7	3	6
Value	3	3	4	2	4

If the maximum weight she can carry is 12 kg, find by trial and error the best combination of items to carry. Use the branch and bound method to confirm your solution as optimal.

3. A small firm orders planks of wood of length 20 m. Each week the firm orders a certain number of planks and then has to meet the orders for that week. Use a bin-filling method to find the minimum number of planks required to meet the following weekly orders.

(a)
Length	Number Required
3 m	5
4 m	6
5 m	2
7 m	2
8 m	1
9 m	1

(b)
Length	Number Required
11 m	1
9 m	1
7 m	2
5 m	2
3 m	12

(c)
Length	Number Required
15 m	1
12 m	2
11 m	1
7 m	3
3 m	2

Does the method always produce the optimum solution?

ANSWERS

The answers to the questions set in the Exercises are given below. Answers to questions set in some of the Activities are also given where appropriate.

1 GRAPHS

Exercise 1A

1. (a) (i) 6 (ii) 4 (iii) 2, 2, 2, 2
   (b) (i) 4 (ii) 6 (iii) 2, 2, 4, 4
   (c) (i) 4 (ii) 5 (iii) 2, 2, 2, 4
   (d) (i) 6 (ii) 15 (iii) 5, 5, 5, 5, 5, 5
   (e) (i) 2 (ii) 1 (iii) 1, 1
   (f) (i) 5 (ii) 4 (iii) 1, 1, 2, 2, 2

2. (i) (a), (d), (e) and (f)
   (ii) (a), (b), (c), (d) and (e)
   (iii) (d) and (e)

3. There are many possible solutions.

Exercise 1B

(a) and (f), (b) and (d), (c) and (k), (g) and (i), (j) and (l) are isomorphic.

Activity 1

Vertices	1	2	3	4	5	6	7	8
Graphs	1	2	4	11	34	156	1044	12346

Exercise 1C

Hints for solution:
1. Any whole number divided by 10 leaves a remainder 0,1, ..., 9.
2. Every whole number except 1 has a smallest prime factor and for the numbers 2 - 20 these come from the set $\{2,3,5,7,11,13,17,19\}$.
4. The coordinates of each lattice point are (odd, odd) or (odd, even) or (even, odd) or (even, even).

Exercise 1D

1. (a) PAQ, PBQ, PCQ, PABQ, PBAQ, PBCQ, PCBQ, PABCQ, and PCBAQ.
   (b) PABPCQ, PABCPBQ, PAQCBQ, ...
   (c) PABPAQ, PABPCBAQ, PAQBAPCQ, ...

(d) PABP, PBCP, PABCP, PAQBP, PAQCP, PBQCP, PABQCP, PBAQCP, PAQBCBP, and their reverses.

2. (a) only

Exercise 1E

(b) and (e) are Eulerian; (d) and (f) are semi-Eulerian.

Activity 7

Vertices	1	2	3	4	5	6	7	8	9	10
Trees	1	1	1	2	3	6	11	23	47	106

Exercise 1G

1. 3
2. 5

Miscellaneous Exercises

2. 5
3. ST, SPT, SQT, SRT, SPQT, SPRT, SQPT, SQRT, SRPT, SRQT, SPQRT, SPRQT, SQRPT, SQPRT, SRPQT and SRQPT.
4. (c) is Eulerian and (a) is semi-Eulerian.
6. 9

2 TRAVEL PROBLEMS

Exercise 2A

1. S-K-L-T = 54; S-L-N-T = 29; S-P-X-T = 16
2. A-Q-R-B = £15
3. M-C-B-D-N = 170 minutes. (Add 10 minutes to each of the journey times not starting at M.)

Exercise 2B

1. 155 m
2. 180 m
3. 500 ECU

Answers

Exercise 2C

1. 36, 63
2. 1700 m
3. 1050 m

Exercise 2D

1. A-B-C-E-D-A = 30
2. E-S-A-F-I-T-A-P-G-E = 917 miles
3. O-B-A-D-C-O = 29 minutes

Exercise 2E

1. 28; 92; 71
2. 660 m
3. 38 km

Miscellaneous Exercises

1. 35
2. 21
3. 108
4. 46
5. 91 km
6. 67 miles
7. 316 miles
8. £250
9. 525 feet
10. Yes - 26 hours travelling plus 48 hours visiting.

3   ITERATION

Exercise 3A

1. $x \approx 1.3$
2. $x \approx 0.7$
3. $x \approx -0.5$, $x \approx 1.4$
4. $x \approx 0.2$, $x \approx 4.5$
5. $x \approx 0.5$, $x \approx 2.6$, $x \approx 4.7$ (twice)

Exercise 3B

1. $x \approx -2.46$
2. $x \approx 1.24$
3. $x \approx -2.86$, $x \approx 2.44$
4. $x \approx 0.590$
5. $x \approx 0.567$

Exercise 3C

$x \approx -1.30$, $x \approx 3.25$

Activity 3

$x \approx -1.16$, $x \approx 1.77$, $x \approx 3.39$

Exercise 3D

1. £553.60
2. 11.1%

Exercise 3E

1. $x \approx 1.46$
2. $x^3 - x^2 - 1 = 0$; $x \approx 1.47$
3. $x \approx 1.31$, $x \approx 0.34$
4. $x_{n+1} = \sqrt[3]{x_n^2 + 5}$; $x \approx 2.12$
5. $x_{n+1} = 2^{\left(\frac{1}{x_n}\right)}$; $x \approx 1.560$

Exercise 3F

1. $x_{n+1} = \sqrt[3]{3x_n + 4}$; $x \approx 2.2$
2. $x_{n+1} = \dfrac{5}{x_n^3} - 2$; $x \approx -2.4$; $x_{n+1} = \left(\dfrac{5 - x_n^4}{2}\right)^{\frac{1}{3}}$; $x \approx 1.2$
3. $x_{n+1} = \dfrac{3^{x_n} - 2}{3}$; $x \approx -0.5$; $x_{n+1} = \dfrac{\ln(3x_n + 2)}{\ln 3}$; $x \approx 1.8$

Exercise 3G

1. $x \approx -1.13$
2. $x \approx 0.519$
3. $x \approx 1.857$; $x \approx 4.536$
4. $f(x) = x^3 - 10$; $x_1 = 2.166.....$; $x_2 = 2.154.....$; 0.003%
5. $x \approx 2.1294$

Miscellaneous Exercises

1. $x \approx 1.521$
2. $x \approx -1.9$ or -1.8; $x \approx -0.3$ or -0.2; $x \approx 2.1$
3. $x \approx -0.42$; $x \approx 3.72$
4. $x \approx -3.68$; $x \approx -2.04$; $x \approx 1.04$; $x \approx 2.68$
5. $x \approx 0.3964$ (the other two are 0.6472 and 1.6708)
6. $x \approx 0.57$

7. $x_{n+1} = \frac{1}{2}\left(x_n + \frac{3}{x_n}\right)$; $x_0 = 1.5$, gives $x \approx 1.732051$

   after just three iterations.

8. $x^4 - 15x^2 - 18x + 81 = 0$; (2.02, 3.45), (3.75, 1.40)

9. 26.58 years

10. $\theta \approx 1.9346$; perimeter ≈ 35.81 cm

4 INEQUALITIES

Activity 2

If $x > y$ and $k < 0$, then $kx < ky$.

Exercise 4A

3. $x \le 2$

4. (a) $\{x \ge 2\}$ (b) $\{-2 \le x < -1\} \cup \{x > 1\}$

5. $\{x \le 1\} \cup \{x \ge 4\}$

Exercise 4B

1. (a) $\{x \le -1\} \cup \{x \ge \frac{2}{3}\}$ (b) $\{-2 \le x \le 4\}$

5. Yes

Activity 4

$A \ge G \ge H$

Exercise 4C

1. (a) A = 2.5   G = 2.21   H = 1.92
   (b) A = 2.5   G = 1.31   H = 0.36
   (c) A = 26.27   G = 2.78   H = 0.37
   (d) A = 251.25   G = 1.57   H = 0.004
   (e) A = 750.00   G = 1.41   H = 0.0027

3. 2

Exercise 4D

1. (a) 9 cm^2 (b) 8 cm^2 (c) 5 cm^2

2. (a) $\frac{\pi}{3\sqrt{3}} \approx 0.605$ (b) $\frac{\sqrt{3}\,\pi}{6} \approx 0.907$

   (c) $\frac{2\pi}{9} \approx 0.698$

3. $\frac{\pi k}{(1+k)^2}$; (a) $k = 1$ (b) $k = 0$

4. $\frac{4\sqrt{3}}{\sqrt{\pi}} \approx 3.909$ cm^3

5. $2\sqrt{2} \approx 2.828$ cm^3

6. Sphere

Miscellaneous Exercises

1. (a) $\left\{-\frac{1}{\sqrt{2}} < x < 0\right\} \cup \left\{x > \frac{1}{\sqrt{2}}\right\}$

   (b) $\left\{-3 < x < -\sqrt{3}\right\} \cup \left\{-1 < x < \sqrt{3}\right\}$

2. $\frac{1}{2} < x < \frac{5}{2}$

3. $\{-1 < x < 1\} \cup \{x > 2\}$

4. $\left\{-\frac{3}{4} < x < 3\right\}$

6. No

8. Equality occurs when numbers are equal.

9. $\dfrac{4\pi\left(k + \dfrac{\pi}{8}\right)}{\left(1 + 2k + \dfrac{\pi}{2}\right)^2}$; $k = \frac{1}{2}$

10. For triangles, I.Q. $\le \dfrac{\pi}{3\sqrt{3}}$, equality only occurring for equilateral triangles.

11. I.Q. ≤ 1, equality only occurs for a sphere.

5 LINEAR PROGRAMMING

Activity 1

Profit is £61; maximum profit of £66 at $x = 6$, $y = 3$.

Exercise 5A

1. If x = no. of blouses, and y = no. of skirts,

   maximise $P = 8x + 6y$

   subject to $x + y \le 7$

   $x + \frac{1}{2}y \le 5$

   $x \ge 0$ and $y \ge 0$.

2. If x = no. of large vans, and y = no. of small vans,

   minimise $P = 40x + 20y$

   subject to $5x + 2y \ge 30$

   $2x + y \le 15$

   $x \le y$

   $x \ge 0$ and $y \ge 0$.

3. If x = no. of boxes of wood screws, and
y = no. of boxes of metal screws,

maximise $P = 10x + 17y$

subject to $3x + 2y \le 3600$

$2x + 8y \le 3600$

$x \ge 0$ and $y \ge 0$.

4. If x = no. of unskilled workers, and
y = no. of skilled workers,

$x + 2y \le 180$

$x + y \ge 110$

$y \ge 40$

$y \ge \frac{1}{2}x$

$x \ge 0$ and $y \ge 0$.

Exercise 5B

1. $x = 3$, $y = 4$ and $P = £48$
2. $x = 4$, $y = 5$ and $P = £260$
3. $P = £18$
   (a) 6 caravans, 6 tents
   (b) 3 caravans, 12 tents
4. 40 adults and 10 junior members
5. $X = \frac{12}{11}$ kg, $Y = \frac{40}{11}$ kg; $X = \frac{4}{7}$ kg, $Y = \frac{30}{7}$ kg

Activity 4

B : $r = t = 0$

C : $s = r = 0$

D : $y = s = 0$

Activity 5

$P = 1600 - \frac{10}{3}s + \frac{160}{3}y$

$P = 2240 + 2s - 160r$

$P = 2320 - 80r - 20t$

Exercise 5C

1. $P = \frac{188}{19}$ at $x = \frac{30}{19}$, $y = \frac{32}{19}$,
2. $P = \frac{43}{11}$ at $x = \frac{28}{11}$, $y = \frac{15}{11}$
3. $P = 56$ at $x = 2$, $y = \frac{12}{5}$

Exercise 5D

1. $P = 21$ at $x = 0$, $y = \frac{7}{2}$
2. $P = 45$ at $x = 0$, $y = \frac{5}{2}$, $z = \frac{15}{8}$

3. $P = 8$ at $x = 0$, $y = 1$, $z = 0$
4. $\frac{12}{5}$ at $x = \frac{2}{5}$, $y = \frac{1}{5}$, $z = 0$

Miscellaneous Exercises

1. Profit = 13860 at $x = 1080$, $y = 180$.
2. Type A = 818, Type B = 164 with profit of £376.40.
3. 19; no. of skilled workers = 12; no. of unskilled workers = 8.
4. No. of small spaces = 48 and no. of large spaces = 24 giving a total 72.
5. 1270 tonnes when 12 Type A and 13 Type B are used.
6. 25 trees; 11 trees
7. Upper seam = $\frac{36}{7}$ h, and lower seam = $\frac{5}{7}$ h.

6 PLANAR GRAPHS

Exercise 6A

1. K_4 is planar; K_6 is non-planar.
2. n odd
3. Tetrahedron, Cube and Octahedron.
4. (a) 10   (b) $\frac{1}{2}n(n-1)$
5. Yes

Exercise 6B

2. rs
4. r and s both even; one even and the other odd

Exercise 6D

1. G_1 and G_3

Miscellaneous Exercises

2. $r > 2$ and $s > 2$
3. (a) 1   (b) 3   (c) 0

7 NETWORK FLOWS

Activity 1

600 cars per hour.

Exrtra capacity from D to A would improve flow.

Exercise 7A

1. Arc AB is over capacity; inflow and outflow are not equal for vertex C.

Exercise 7B

1. 16 (A - E)
2. (a) 26 (b) 16 (c) 53

Exercise 7C

1. 18, 57, 38
2. 75

Exercise 7D

1. (a) 28 (b) 23 (c) 111
2. 16

Exercise 7E

1. 19
2. No; No
3. 19

Miscellaneous Exercises

1. $N_1 - 30$; $N_2 - 15$; $N_3 - 29$
2. 46
3. (a) No (b) Yes (c) No (d) Yes;
   18 for (b) and (d)
4. 42
5. 5200
6. 1800
7. 8
8. $8 \times 500 = 4000$

8 LOGIC

Activity 2

PM	MP	PM
SM	MS	MS
S P	S P	S P

Activity 3

(a) F (b) V (c) F (d) V

Exercise 8A

1. (a) $p \wedge q$ (b) $\sim p \wedge \sim q$ (c) $p \oplus q$
   (d) $\sim(p \wedge q)$
2. (a) The cooker is working but the visitors are not hungry.
   (b) There is enough food and the visitors are hungry but the cooker is not working.
   (c) Either the visitors are hungry or there is not enough food.
   (d) Either the cooker is working and there is enough food or the visitors are not hungry.
   (e) There is not enough food and either the cooker is not working or the visitors are not hungry.

Exercise 8B

1.
q	r	$q \vee r$
0	0	0
0	1	1
1	0	1
1	1	1

2.
p	r	$\sim p \wedge r$
0	0	0
0	1	1
1	0	0
1	1	0

3.
p	r	$p \vee \sim r$
0	0	1
0	1	0
1	0	1
1	1	1

4.
p	q	$\sim p \vee \sim q$
0	0	1
0	1	1
1	0	1
1	1	0

Exercise 8C

1.
a	b	c	$(a \vee b) \vee c$
0	0	0	0
0	0	1	1
0	1	0	1
0	1	1	1
1	0	1	1
1	0	1	1
1	1	0	1
1	1	1	1

2.
a	b	c	$a \wedge (b \wedge c)$
0	0	0	0
0	0	1	0
0	1	0	0
0	1	1	0
1	0	0	0
1	0	1	0
1	1	0	0
1	1	1	1

3.
a	b	c	$a \vee (b \vee c)$
0	0	0	0
0	0	1	1
0	1	0	1
0	1	1	1
1	0	0	1
1	0	1	1
1	1	0	1
1	1	1	1

4.
a	b	c	$(a \wedge b) \wedge c$
0	0	0	0
0	0	1	0
0	1	0	0
0	1	1	0
1	0	0	0
1	0	1	0
1	1	0	0
1	1	1	1

5.
a	b	c	$a \wedge (b \vee c)$
0	0	0	0
0	0	1	0
0	1	0	0
0	1	1	0
1	0	0	0
1	0	1	1
1	1	0	1
1	1	1	1

6.
a	b	c	$(a \wedge b) \vee (a \wedge c)$
0	0	0	0
0	0	1	0
0	1	0	0
0	1	1	0
1	0	0	0
1	0	1	1
1	1	0	1
1	1	1	1

7.

a	b	c	a∨(b∧c)
0	0	0	0
0	0	1	0
0	1	0	0
0	1	1	1
1	0	0	1
1	0	1	1
1	1	0	1
1	1	1	1

8.

a	b	c	(a∨b)∧(a∧c)
0	0	0	0
0	0	1	0
0	1	0	0
0	1	1	1
1	0	0	1
1	0	1	1
1	1	0	1
1	1	1	1

Activity 6

a	b	a⇒b
0	0	1
0	1	1
1	0	0
1	1	1

Exercise 8D

1. (a) 1 (b) 1 (c) 0 (d) 0 (e) 1
2. (a) b⇒a (b) ~c∧~b∧~a (c) c⇒a
    (d) b∧~c⇒a (e) ~b⇒c∧a
3. (a) Either I water the plants and the crops grow or I spread manure and the crops grow.
    (b) If I either spread manure or do not water the plants, then the crops do not grow.
    (c) If the crops grow then I water the plants and spread manure.
    (d) If the crops do not grow or I spread the manure then I water the plants.

Exercise 8E

1. (a) T (b) F
2. (a) If and only if the theme park has excellent rides and the entrance charges are not high, then attendances are large.
    (b) If the attendances are not large or the entrance charges are not high, then the theme park has excellent rides.

Exercise 8F

1. Contradiction
2. Contradiction
3. Tautology

Exercise 8G

1. No
2. No
3. Yes
4. No

Miscellaneous Exercises

1. (a) a∧~b (b) a∧b (c) a∧b
    (d) ~a∧~b (e) a⇒b (f) ~b⇒~a
    (g) a∧~b (h) ~b⇒a

2. (a)

p	q	(p∨~q)⇒q
0	0	0
0	1	1
1	0	0
1	1	1

(b)

p	q	[p∨(~p∨q)]∨(~p∧~q)
0	0	1
0	1	1
1	0	1
1	1	1

(c)

p	q	(~p∨~q)⇒(p∧~q)
0	0	0
0	1	0
1	0	1
1	1	1

(d)

p	q	~p⇔q
0	0	0
1	0	1
0	1	1
1	1	0

(e)

p	q	r	(~p∧q)∨(r∧p)
0	0	0	0
0	0	1	0
0	1	0	1
0	1	1	1
1	0	0	0
1	0	1	1
1	1	0	0
1	1	1	1

(f)

p	q	(p⇔q)⇒(~p∧q)
0	0	0
0	1	1
1	0	1
1	1	0

3. (a) No (b) No (c) Yes (d) Yes
4. (a) No (b) Yes (c) No (d) No
5. (a) Yes (b) No (c) No (d) No
6. (a) Yes (b) No (c) Yes

9 BOOLEAN ALGEBRA

Exercise 9A

1.

x	y	$(\sim y \vee x)$	$x \wedge (\sim y \vee x)$
0	0	1	0
0	1	0	0
1	0	1	1
1	1	1	1

2.

a	b	c	$(\sim b \wedge c)$	$a \vee (\sim b \wedge c)$
0	0	0	0	0
0	0	1	1	1
0	1	0	0	0
0	1	1	0	0
1	0	0	0	1
1	0	1	1	1
1	1	0	0	1
1	1	1	0	1

3.

a	b	c	$(\sim b \vee c)$	$a \vee (\sim b \vee c)$	$[a \vee (\sim b \vee c)] \wedge \sim b$
0	0	0	1	1	1
0	0	1	1	1	1
0	1	0	0	0	0
0	1	1	1	1	0
1	0	0	1	1	1
1	0	1	1	1	1
1	1	0	0	1	0
1	1	1	1	1	0

4. $(a \wedge b) \vee \sim c$

5. $[\sim(p \wedge q) \vee (p \wedge r)] \wedge \sim r$

Exercise 9B

1. Equivalent
2. Equivalent
3. Equivalent
4. Not equivalent

Exercise 9C

1. $[A \wedge (B \vee D)] \vee (\sim D \wedge C)$

2. $A \wedge [(A \vee D \vee C) \vee (C \wedge (B \vee \sim D))]$

Exercise 9D

1. $a \vee b$
2. $a \wedge b$
3. $(a \wedge b) \vee (c \wedge d)$

Exercise 9E

1. $f(a,b) = (\sim a \wedge \sim b) \vee (a \wedge \sim b)$

2. $f(a,b) = (\sim a \wedge \sim b) \vee (\sim a \wedge b) \vee (a \wedge b)$

3. $f(x,y,z) = (\sim x \wedge \sim y \wedge \sim z) \vee (x \wedge \sim y \wedge \sim z) \vee (x \wedge y \wedge z)$

Exercise 9F

1. $a \wedge b = (\sim a \uparrow b) \uparrow (a \uparrow b)$

2. $a \wedge \sim b = [a \uparrow (b \uparrow b)] \uparrow [a \uparrow (b \uparrow b)]$

3. $(\sim a \wedge \sim b) \vee \sim b = [(a \uparrow b) \uparrow (a \uparrow b)] \uparrow [(b \uparrow b) \uparrow (b \uparrow b)]$

Activity 4

a	b	c	Carry bit	Answer bit
0	0	0	0	0
0	0	1	0	1
0	1	0	0	1
0	1	1	1	0
1	0	0	0	1
1	0	1	1	0
1	1	0	1	0
1	1	1	1	1

Miscellaneous Exercises

2. (a) $(\sim p \vee q) \wedge \sim r$

   (b) $(\sim q \wedge r) \vee p$

   (c) $[(\sim q \vee r) \wedge p] \vee q$

3. (a) $(A \vee B \vee C \vee \sim B) \wedge C$

   (b) $[(P \vee Q) \wedge (R \vee P \vee S)] \wedge (\sim Q \wedge \sim P)$

   (c) $(A \wedge D) \vee (B \wedge E) \vee (A \wedge C \wedge E) \vee (B \wedge C \wedge D)$

8. (a) $b \wedge c$ (b) $a \vee b$ (c) $(p \wedge q) \vee r \vee s$

9. $(a \wedge b) \vee c \wedge \{c \wedge [(\sim a \wedge b) \vee (a \wedge \sim b)]\}$

10. $[(a \uparrow a) \uparrow (a \uparrow a)] \uparrow (b \uparrow b)$

10 DIFFERENCE EQUATIONS 1

Exercise 10A

1. (a) 11   (b) 58   (c) 12

2. (a) $u_n = u_{n-1} + 2$, $u_1 = 3$, $n \geq 2$

   (b) $u_n = 2u_{n-1} + 1$, $u_1 = 2$, $n \geq 2$

   (c) $u_n = 3u_{n-1} - 1$, $u_1 = 1$, $n \geq 2$

3. $u_n = \frac{1}{3}u_{n-1}$; between 12 and 13 strokes.

4. (a) 1, 4, 9, 16; $u_{n+1} = u_n + 2n + 1$, $n \geq 1$

   (b) 9, 17, 24, 30; $u_{n+1} = u_n + 9 - n$, $n \geq 1$

   (c) 9, 24, 45, 72; $u_{n+1} = u_n + 3(2n+3)$, $n \geq 1$

Exercise 10B

1. (a) $u_n = u_1 + 2(n-1)$

   (b) $u_n = 4^{n-1}u_1 + \frac{1}{3}\left(1 - 4^{n-1}\right)$

   (c) $u_n = 3^{n-1}u_1 + 3^{n-1} - 1$

2. $p_n = 1.025p_{n-1}$; $p_n = (1.025)^n p_0$;
   $p_{20} \approx 819$ million; 16 - 17 years.

3. £24.96

Exercise 10C

1. (a) $u_n = 4^{n-1}u_1 + \frac{2}{3}\left(4^{n-1} - 1\right)$

   (b) $u_n = 4^n u_0 + \frac{2}{3}\left(4^n - 1\right)$

   (c) $u_n = 3^n u_0 - \frac{5}{2}\left(3^n - 1\right)$

   (d) $u_n = u_0 + 6n$

   (e) $u_n = u_1 - 8(n-1)$

   (f) $u_n = (-2)^n u_0 - \frac{4}{3}\left((-2)^n - 1\right)$

   (g) $u_n = (-3)^n u_0 - \frac{1}{2}\left((-3)^n - 1\right)$

   (h) $u_n = (-4)^n u_0 + \frac{3}{5}\left((-4)^n - 1\right)$

   (i) $u_n = 4^{n-1}u_1$

   (j) $u_n = 4^n u_0 - \frac{5}{3}\left(4^n - 1\right)$

2. (a) $u_n = \frac{1}{2}\left(7 \times 3^n - 5\right)$

   (b) $u_n = 2 + (-2)^{n-1}$

   (c) $u_n = 4 - 3n$

   (d) $u_n = \frac{3}{4}\left(5^{n-1} - 1\right)$

   (e) $u_n = 3 + 7n$

   (g) $u_n = \frac{3}{4}(-3)^n + \frac{1}{4}$

Exercise 10D

1. (a) $u_n = 3^{n-1}u_1 + 2\left(3^{n-1} - 1\right)$

   (b) $u_n = \left(\frac{1}{2}\right)^{n-1}u_1 - \left(\frac{1}{2}\right)^{n-3} + 4$

   (c) $q_n = 3n - 2$

   (d) $a_n = 2^{n+1}$

   (e) $b_n = 2 \times 4^{n-1} + \frac{5}{3}\left(4^{n-1} - 1\right)$

2. $u_n = 7 \times 2^{n-1} + 3\left(2^{n-1} - 1\right)$, $n \geq 1$

3. £24.95

4. £70.51

6. 6418 tonnes; 16086 tonnes

7. Approximately 26 months

Exercise 10E

1. $u_n = \frac{1}{2}(n+2)(n-1) + 5$

2. (a) $u_n = u_1 + \frac{n}{6}(n+1)(2n+1) - 1$

   (b) $u_n = u_1 + 2^2\left(2^{n-1} - 1\right)$

   (c) $u_n = 2^{n-1}u_1 + 3 \times 2^{n-1} - (n+2)$

3. $k = 3$; $u_6 = 1577$

4. 126%

5. £40.47

Miscellaneous Exercises

1. (a) $u_n = 2^{n-1}u_1$ (b) $u_n = 3^{n-1}u_1 + \frac{3}{2}\left(3^{n-1} - 1\right)$

   (c) $u_n = 3^{n-1}u_1 + \frac{5}{4} \times 3^{n-1} - \frac{1}{2}n - \frac{3}{4}$

2. $u_n = 3u_{n-1} - 2$; $u_n = 3^{n-1} + 1$; $u_{10} = 19684$

3. £60.28

4. 16.35 million

5. n; $u_n = u_{n-1} + n - 1$; $u_n = \frac{n(n-1)}{2}$; $u_{20} = 190$

6. $p = 4$, $q = -5$, $u_6 = 343$

7. (a) £315.96   (b) £300

8. 47 million, 37 million

9. 47 months

10. Yes

11 DIFFERENCE EQUATIONS 2

Exercise 11A

1. 12, 29, 70

2. $p = 1$, $q = 6$

3. $+1$ for n odd, -1 for n even

4. Repeats after six terms

5. $\frac{1}{2}\left(1+\sqrt{5}\right)$ (golden ratio)

6. 2

Exercise 11B

1. (a) $u_n = A3^n + B(-2)^n$

   (b) $u_n = A\left(2+\sqrt{5}\right)^n + B\left(2-\sqrt{5}\right)^n$

   (c) $u_n = A2^n + B(-1)^n$

2. $u_n = \frac{1}{\sqrt{5}}\left\{\left(\frac{1+\sqrt{5}}{2}\right)^{n+1} - \left(\frac{1-\sqrt{5}}{2}\right)^{n+1}\right\}$ $(n = 0,1,...)$

3. $u_n = \frac{1}{2}(1-i)(2i)^n + \frac{1}{2}(1+i)(-2i)^n$

4. $u_n = 4^n + 3\times 2^n$; $u_6 = 4288$

5. $u_n = u_{n-1} + 6u_{n-2}$; $u_n = 3^n + 3\times(-2)^n$

Activity 3

If $u_n = $ no. of pairs of mice; $u_n = u_{n-1} + 2u_{n-2}$

If $u_1 = u_2 = 10$, then $u_{12} = 13650$

Exercise 11C

1. (a) $u_n = A2^n + Bn2^n$

   (b) $u_n = A + Bn$

2. $u_n = 3^n(2+n)$

3. $u_n = 2\left(i^n + (-i)^n\right)$

4. $u_n = (-1)^n(4-3n)$

5. (a) $u_n = 2\times 5^n + 3\times(-3)^n$

   (b) $u_n = \frac{1}{2}\left(\left(\sqrt{3}\right)^n + \left(-\sqrt{3}\right)^n\right)$

   (c) $u_n = 3^n(1+2n)$

6. $u_n = 2\left(u_{n-1} + u_{n-2}\right)$, $u_0 = u_1 = 1$;

   $u_n = \frac{1}{2}\left(\left(1+\sqrt{3}\right)^n + \left(1-\sqrt{3}\right)^n\right)$

Activity 4

$u_n = -\frac{81}{4}\left(\frac{1}{3}\right)^n + 16\left(\frac{1}{2}\right)^n - \frac{3}{4} + \frac{1}{2}n$

Exercise 11D

1. $u_n = y_n + v_n$ where $y_n = A2^n + B3^n$ and

   (a) $v_n = 1$

   (b) $v_n = \frac{1}{4}(7+2n)$

   (c) $v_n = \frac{1}{2}n^2 + \frac{7}{2}n + 8$

   (d) $v_n = \frac{1}{6}\times 5^{n+2}$

   (e) $v_n = -2n2^n$

2. $u_n = \frac{1}{2}\times 4^n - \frac{5}{3}\times 3^n + 2^{n+1}$

3. $u_n = \frac{1}{49}\left(29\times(-5)^{n-1} + 41\times 2^{n-1}\right) + \frac{2}{7}n2^n$

Exercise 11E

1. (a) $\dfrac{x}{1-2x-8x^2}$ (b) $\dfrac{2+7x}{1+x-3x^2}$

   (c) $\dfrac{1+3x}{1-4x^2}$ (d) $\dfrac{1}{1-2x}$

2. (a) $\dfrac{1}{x-3} + \dfrac{2}{x+1}$ (b) $\dfrac{2}{2x-5} - \dfrac{1}{x-2}$

   (c) $\dfrac{4}{x-3} - \dfrac{3}{x+3}$

3. (a) $1 + x + x^2 + ... + x^n + ...$

   (b) $1 + 2x + 4x^2 + ... + 2^n x^n + ...$

   (c) $1 - 3x + 9x^2 + ... + (-1)^n 3^n x^n + ...$

   (d) $1 + 2x + 3x^2 + ... + (n+1)x^n + ...$

   (e) $3\left(1 - 4x + 12x^2 + ... + (-1)^n(n+1)2^n x^n + ...\right)$

4. (a) $u_n = 4^{n+1} - 4(-1)^n$

   (b) $u_n = 3\times 4^n$

5. $G(x) = \dfrac{1}{1-x-x^2}$

6. $u_n = 2\times 3^n + 3\times(-3)^n$

Exercise 11F

1. (a) $\dfrac{2}{1-2x} - \dfrac{1}{1-x}$ (b) $\dfrac{1}{3(2-x)} - \dfrac{5}{3(1+x)}$

   (c) $\dfrac{1}{3(x+1)} + \dfrac{1}{6(1-2x)} + \dfrac{3}{2(1-2x)^2}$

2. (a) $\dfrac{1}{1-x}$ (b) $\dfrac{1}{(1-x)^2}$ (c) $\dfrac{-1}{x(1-x)^2}+\dfrac{1}{x}$

   (d) $\dfrac{x^2}{(1-x)^2}$ (e) $\dfrac{1}{(1-2x)}$ (f) $\dfrac{25x^2}{(1-5x)}$

3. (a) $1+3x+9x^2+\ldots+3^n x^n+\ldots$

   (b) $\dfrac{1}{4}+\dfrac{x}{4}+\dfrac{3x^2}{16}+\ldots+\dfrac{(n+1)x^n}{2^{n+2}}+\ldots$

   (c) $-x-x^2-x^3-\ldots+(-x^n)+\ldots$

4. (a) $u_n=\tfrac{1}{9}\left(7\times4^n+(-2)^n-8\right)$

   (b) $u_n=3^{n+1}-2^{n+1}$

   (c) $u_n=4\times2^n-\tfrac{1}{2}n^2-\tfrac{5}{2}n-4$

Miscellaneous Exercises

1. $u_n=3\times2^n+2\times(-2)^n$

2. $u_n=(A+Bn)2^n$

3. Fibonacci Sequence

4. (a) $u_n=-1+3\times2^{n-1}$

   (b) $u_n=-n-1+4\times2^{n-1}$

   (c) $u_n=\dfrac{1}{2\sqrt{3}}\left((1+\sqrt{3})^n-(1-\sqrt{3})^n\right)$

5. $u_n=-\tfrac{2}{5}\times2^n+\tfrac{3}{7}\times4^n-\tfrac{1}{35}\times(-3)^n$

6. $N_t=3-\tfrac{27}{8}\left(\tfrac{4}{3}\right)^t+\tfrac{5}{4}\times2^t$

7. $a=2,\ b=-3,\ k=9;\ u_n=\tfrac{1}{16}\left(-1+(-3)^n\right)+\tfrac{9}{4}n$

8. $u_n=5(3^n-1);\ n=12$

9. $u_n=\tfrac{5}{8}+\tfrac{3}{8}\times(-1)^n+\tfrac{1}{4}n^2+\tfrac{1}{2}n$

12 CRITICAL PATH ANALYSIS

Activity 1

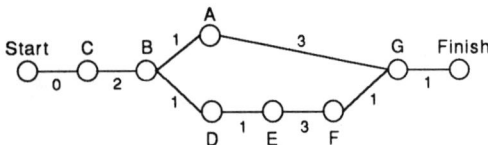

Exercise 12A

1. See Activity 1 above.

2.

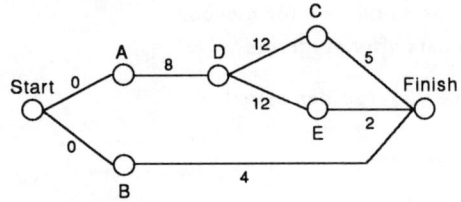

3.

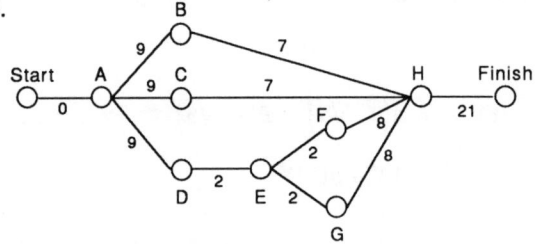

4.

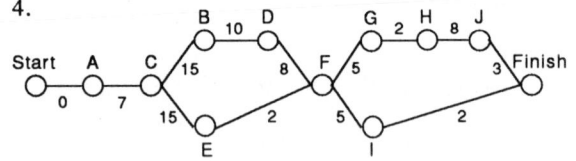

Activity 2

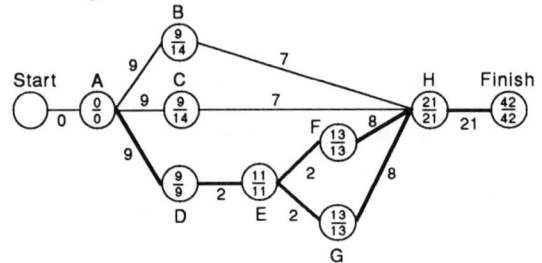

Critical path is shown by **bold** lines in the network.

Exercise 12B

1. (a) Start - A - B - C - F - I - Finish
   (b) Start - A - B - D - E - G - Finish
   (c) Start - A - C - E - G - F - Finish

2. Start - A - C - B - D - F - G - H - J - Finish

Miscellaneous Exercises

1. (a) (i) 13 (ii) start - C - F - Finish
   (b) 2

2. (b) Start - B - C- D - G - H- Finish

(c)

Activity	Latest Starting Time
A	8
B	0
C	4
D	12
E	22
F	15
G	18
H	28
I	16
J	20

3. (b) Start - A - D - H - Finish

(c)

	Starting Times	
Activity	Earliest	Latest
A	0	0
B	4	5
C	0	1
D	4	4
E	0	4
F	4	13
G	4	8
H	9	9

4. (b) and (c)

	Starting Times		
Activity	Earliest	Latest	Float
A	9	15	6
B	12	13	1
C	0	0	0 ←
D	18	19	1
E	9	9	0 ←
F	0	4	4
G	21	22	1
H	7	10	3
I	12	12	0 ←
J	0	2	2
K	21	21	0 ←

Critical path: Start - C - E - I - K - Finish

5. (b) 75 minutes

(c) Start - A - B - C - E - F - G - I - Finish

13 SCHEDULING

Activity 1

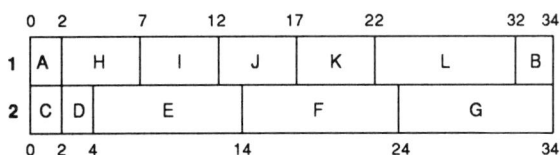

Activity 2

Activity	Rank
A	0
B	15
C	15
D	15
E	7
F	7
G	7
H	4
I	4
J	4
K	4
L	14

This method gives a finishing time of $t = 24$, which is **not** optimal.

Exercise 13A

1.

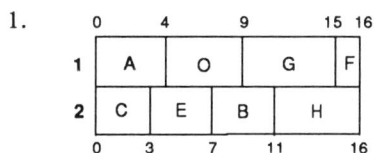

Optimum solution is produced.

2.

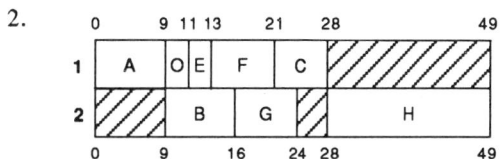

Similar solution for second method.

These are optimal solutions for 2 workers.

3. For the problem in Activity 1, this method gives a completion time of 25 using 4 workers. It is not optimal.

Exercise 13B

1. (a)

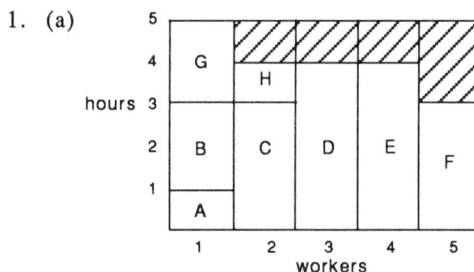

(b) 4, 4, 3, 3, 2, 2, 1, 1

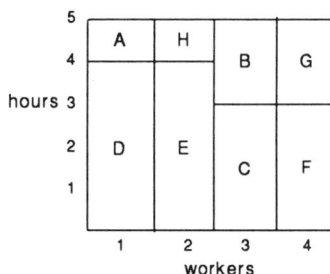

2.

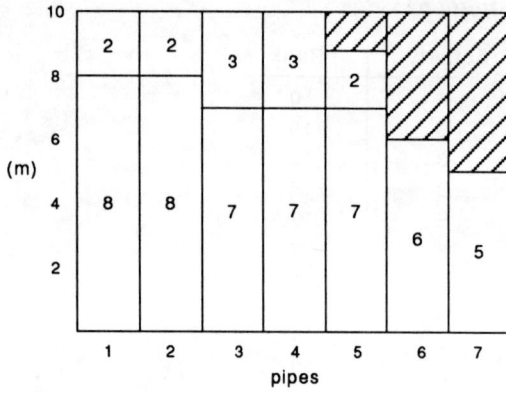

3. It is possible to meet the order with just 10 sheets, cut appropriately.

Exercise 13C

1. C and E, giving value 14.
2. A and C, giving savings of £215 000.
3. C, D and E, giving value 18.

Miscellaneous Exercises

1. (a) 4 workers   (b) (i) 6   (ii) 5, yes
2. A, B, D, giving value 8.
3. Optimum solutions:
   (a) 4   (b) 4   (c) 5

FURTHER READING

As this is a new topic for schools and colleges, there are unfortunately few relevant resources. Most books on Decision Mathematics (or Discrete Mathematics) have been written for students taking courses in higher education. The list below gives the most relevant texts, but none of them cover all the material in the AEB syllabus at a suitable depth.

1.   **Decision Mathematics,** 2nd edition, *The Spode Group* , 1986 (Ellis Horwood) 0 13 200973 0

2.   **Decision Mathematics,** Vol. 1 and 2, *R. Davison* and *L. Cochrane,* 1991 (Cranfield) 1 871315 30 1, 1 871315 32 8

3.   **Graphs, Networks and Design,** Open University Course TM361, 1981

4.   **Essential Discrete Mathematics,** *R. Johnsonbaugh,* 1986 (Macmillan) 0 02 360630 4

5.   **Discrete Mathematics,** *R. Johnsonbaugh,* 1984 (Macmillan)   0 02 360900 1

6.   **Discrete Mathematics,** *Norman L. Biggs,* 1989 (OUP) 0 198 53252 0

7.   **Modern Analytical Techniques,** 2nd edition,   *F. Owen* and *R. Jones,* 1984 (Polytech) 0 85505 081 0

8.   **Decision Making, Models and Algorithms,** *Saul I. Gass* 1985 (John Wiley & Sons) 0 471 80963 2

9.   **Model Building in Mathematical Programming,** 2nd edition, *H. Williams,* 1984 (John Wiley & Sons) 0 471 90606 9

10. **Discrete Mathematics for New Technology,** *R. Garnier* and *J. Taylor,* 1992 (Adam Hilger) 0 7503 0136 8

INDEX

Accuracy in iteration 42
Activity network 194
Adder circuits 135
Affirmation 118
Algebra, Boolean 135
Algorithm 18
    activity network 196
    critical path analysis 193
    first fit 213
    first fit decreasing 213
    greedy 23
    Kruskal's 22
    labelling procedure 107
    minimum connector 21
    planarity 97
    Prim's 222
    shortest path 17
    simplex 87
    scheduling 207
    travelling salesman problem 27
AND gate 117
Arc 1
Argument 118
Arithmetic mean 69
Arithmetic progression 166
Associative laws 142
Augmented matrix 92
Auxiliary equation 175

Bin packing 211
Binomial theorem 191
Bipartite graph 99
Bisection 44
Bit 135
Boole, George 123, 135
Boolean algebra 135
    associative laws 142
    complemet laws 143
    commutative laws 142
    de Morgan's laws 142
    distiributive laws 142
    gates 136
    identity laws 142
    truth tables 136
Boolean expression 123, 144
Boolean function 144
Bounds for travelling salesman problem 30

Branch and bound method 265

Capacity 108
Carroll, Lewis 117
Circuits 139
    adder 135
    closed 139
    combinatorial 135
    full adder 148
    half adder 148
    open 139
    switching 139
Chinese postman problem 32
Commutative laws 142
Complement laws 143
Complete graph 2, 15
Compound propositions 121
Conjection 121
Connected graphs 2, 15
Connectives 121
Constraints 80
Contradictions 129
Convergence 53
Critical path analysis 193
Cubes 13
Cut 109
Cycle 8

de Morgan's laws 147
Degree 2
Difference equations
    auxiliary equation 175
    first order 153
    iteration 157
    linear first order 158
    linear second order 175
    general solution 174
    generating function 187
    homogeneous 166, 175
    non-homogeneous 182
    particular solution 182
    second order 177
Digraphs 108
Disconnected graph 2
Disjunctive normal form 146, 152
Distributive laws 142

Earliest start time   200
Edges   1, 108
Equivalence   128
Eulerian graph   33, 98
Excluded region   63
Exclusive   122

Feasible region   84
Ferrari, Ludovico   37
del Ferro, Scipione   37
Fibonacci   174
First fit decreasing method   212
First fit packing   211
Float   202
Flow
    augmenting path   115
    maximum   109
    network   108
    value   109
Full adder   148

Galois, Evariste   40
Game strategies   12
General solution   174
Geometric mean   69
Generating function   187
Graph
    complete   2, 15
    connected   2, 15
    cycles   8
    degree   2
    disconnected   2
    edges   1
    Eulerian   8
    Hamiltonian   9
    handshake lemma   6
    path   7
    Petersen   103
    planar   97
    simple   2
    subgraph   2
    trail   7
    trees   10
    vertices   1
    walk   7
Graphical methods   39
Greedy algorithm   23

Half adder   148
Hamiltonian cycle   9, 28
Handshake lemma   6
Harmonic mean   70
Homogeneous difference equations   166, 175

Icosian game   9
Identity laws   142
Implications   126
Inclusion   122
Inequalities   63
    linear   65
    graphs   66
    isoperimetric   72
Interpolation   48
Interval bisection   44
Isomorphism   1, 3
Isoperimetric quotient number   74
Iteration   37
    accuracy   42
    graphical methods   39
    internal bisection   44
    locating root   46
    linear interpolation   48
    rearrangement methods   51
    convergence   53
    Newton's method   54
    secant method   57
    difference eqations   156

Knapsack problem   215
Konigsberg bridges   8
Kruskal's algorithm   22
Kuratowski's theorem   104

Labelling flows   107
Latest starting time   200
Linear difference equation   158, 175
Linear interpolation   48
Linear programming   79
Loans   163
Logic   117
    affirmation   118
    argument   131
    connectives   121
    conjection   121
    contradiction   129
    disjunction   122
    equivalence   128
    exclusive   122
    gates   136
    implication   126
    inclusive   122
    negation   121
    predicate   118
    proposition   118
    syllogisms   118
    tautologies   129
    truth tables   126

Locating root 46
Loop 2
Lower bound 30

Maximum capacity 114
Maximum flow 109
Means 69
Minimum cut 109
Minimum connector problem 21
de Morgan's laws 147

NAND gate 147
Negation 121
Network
    activity 194
    capacity 109
    critical path 193
    minimum cut 109
    flows 107
    values 109
Newton's method 54
NOT gate 136

OR gate 122

Path 7
Petersen graph 103
Pigeon-hole principle 6
Planar graphs 97
Planarity algorithm 100
Plane drawings 98
Populations 167
Precedence relations 194
Predicate 118
Prims' algorithm 22
Propositions 118

Rearrangement methods 51
Recursion 154
Roots 46

Scheduling 208
Secant method 57
Second order difference equation 173
Semi-Eulerian 9
Shortest path 17
Simple graph 2
Simplex method 87
Simplex tableau 91
Sink 108
Slack variables 88
Source 108

Spanning tree 21
Subdivision 104
Subgraph 2, 104
Subject 118
Super Sources and sinks 113
Switching circuits 139
Syllogisms 118

Tautologies 129
Theorem
    Kuratowski's 104
    maximum flow - minimu cut 109
Tower of Hanoi 153
Travelling salesman problem 27
Tree 10
Truth tables 124

Upper bound 30

Valid argument 131
Value 109
Vertices 1

Walk 7